STRUCTURE AND PROPERTIES
OF MAGNETIC MATERIALS
D.J.Craik

Applied physics series

Series editor H.J.Goldsmid

STRUCTURE AND PROPERTIES
OF MAGNETIC MATERIALS
D.J.Craik

Pion Limited, 207 Brondesbury Park, London NW2 5JN

Library edition   SBN 85086 017 2
Student edition   SBN 85086 018 0

Set on IBM 72 Composers by Pion Limited, London.
Printed in Great Britain by J.W.Arrowsmith Limited, Bristol.

Contents

Preface

This is an introductory text intended for those largely unfamiliar with the principles relating to the structure and behaviour of magnetic materials. As such it aims to commence each chapter or section at an elementary level, to introduce or refresh the reader's memory on the mathematics involved, and to proceed to a level at which the principles can just be meaningfully understood. As there are always several aspects which must be covered, in some respects this approach leads to a certain limitation of scope: particularly perhaps in giving the simple account of crystal fields without the introduction of covalency and also in the rather limited treatment of exchange effects. However, as this is an introductory text it has seemed better to err on the side of simplicity and the reader for whom this account is insufficient should proceed to the major works (such as the excellent, though rather forbidding, *Magnetism* series of volumes edited by G. T. Rado and H. Suhl and published over the last few years by the Academic Press) with, it is hoped, a good overall grasp of the subject.

In view of the existing literature the main requirement seemed to be for a compact and economical work, suitable for students in various disciplines and also for research workers in a simple introductory sense, dealing with the *behaviour* of magnetic materials as affected by their microstructure and such factors as crystal size and shape. However, it is impossible to fully appreciate this behaviour without some under-standing of the origins of the intrinsic properties, such as spontaneous magnetization and magnetocrystalline anisotropy, which in turn arise from the atomic structure and interactions within the crystal structure. The production of the book would appear to be most timely, since there certainly seems to be a strong resurgence of interest in magnetic behaviour: current technological requirements have brought home the realisation that many long-recognised problems still require considerable elucidation. Professor Néel (in the light of whose recent Nobel prize the least worthy workers must to some extent bask) has clearly stressed the necessity for further study of such basic and simply-stated problems of magnetic behaviour as coercivity.

The magnetic dipole is the entity most basic to the study of magnetic materials and after introducing the magnetostatics from the obvious standpoint of the effects of charges in motion, magnetic materials are described and classified in terms of assemblies of dipoles. The magneto-statics is devoted substantially to the elucidation of energy terms, while the second chapter deals mainly with the statics and dynamics of dipoles in applied fields and in effective fields. The 'field' and 'induction' terminology is used rather than 'applied flux density' or 'applied induction', and 'induced flux density' or 'resulting induction', largely as a matter of convenience and clarity. There is no need to connect different terminologies with different systems of units. The present work follows the great bulk of the research literature in using c.g.s. units but includes a table as an Appendix which allows conversions to SI units [Système International: see D. H. Smith, *Contemporary Physics*, **11**, 287 (1970); L. F. Bates, *ibid.*, **11**, 301 (1970)]. It is important to note that two different SI conventions exist, one which uses the relation between induction, field and magnetization

$$B = \mu\mu_0 H = \mu_0(H+M) \qquad (\mu = 1+M/H)$$

and one which uses a relation between induction and a quantity J or I which is sometimes called the magnetization [H. Zijlstra, *Experimental Methods in Magnetism* (North Holland, Amsterdam, 1967)] and sometimes the magnetic polarisation

$$B = \mu\mu_0 H = \mu_0 H + J \qquad (\mu = 1+J/\mu_0 H) \, .$$

$\mu_0 = 4\pi \times 10^{-7}$ henry per metre so, while B and H always have different dimensions, B and M have different dimensions according to the first convention but B and J have the same dimensions according to the second. Again, H and M have the same dimensions whereas H and J do not. The first convention is the one accepted by the International Union of Pure and Applied Physics and is embodied in the 1969 Report of the Royal Society, Symbols, Signs, and Abbreviations: the unit of magnetic field is 1 ampere per metre and that of induction or flux density is the tesla or volt second per square metre. The most important conversions, therefore, are effected as $1 \text{ A m}^{-1} = 4\pi \times 10^{-3}$ Oe (oersted, c.g.s.) and $1 \text{ T} = 10^4$ G (gauss: c.g.s.). The characteristic B versus H, and $(B-H)$ versus H, or $4\pi I$ versus H, or I versus H, graphs or loops can readily be converted from one system to the other. A magnetization loop or curve would, according to the first SI system give both M and H in units of A m^{-1} while a J versus H loop (on the second system) would consist of teslas against A m^{-1}. Alternatively, some authors may plot applied induction B_0 against the measured induction B, both in teslas. From this one may obtain a plot of $B-B_0 = J$ against B_0.

Returning to the theme of the monograph, it is next clearly necessary to describe the origin of the arbitrary dipoles in real materials. This is done in two stages—first by considering the magnetic dipole moments of free or isolated atoms and ions, and then proceeding to the important modifications due to their incorporation into crystals. Thus Chapter 3 deals with atomic structure from first principles, introducing the simple quantum theory needed, while Chapter 4 introduces crystal field effects, exchange and anisotropy. This manner of presentation is most pertinent to the ionic magnetic materials which are of such current technical importance and theoretical interest and in which the 'magnetic electrons' can be considered as basically localized. The treatment of metals follows a trifle arbitrarily; had they been the chief concern one would have commenced with free electron theory and band theory.

After discussion of the origin of the intrinsic properties of magnetic materials (principally the spontaneous magnetization, exchange energy and magneto-crystalline anisotropy energy), the behaviour of materials from which domains are specifically precluded and where only magnetization rotation is permitted is described in Chapter 5. The main parameters influencing such behaviour in practice are crystal size and shape.

An extensive treatment of domain structures in Chapter 6, which is also a little more mathematically complex, is justified by their relevance to important technical problems and by a current resurgence of interest in this aspect of magnetism. It is hoped that Chapters 5 and 6 in particular will interest those who wish to understand the principles underlying the properties and applications of magnetic materials in the fields of technology: Bates (*loco cit*) has referred, with some justification, to "ferromagnetic materials on whose exploitation our civilization depends".

Apart from serving as an introductory text it is naturally hoped that this work in itself will prove adequate for postgraduate courses in solid state physics, materials science, etc. and for undergraduate courses in which magnetism is given some prominence.

The author is indebted to those who have given permission for their results to be quoted, particularly to Philips Research Laboratories, Eindhoven, for the micrographs of Figures 5.12 and 6.12a, to C.N.R.S., Bellevue, for 6.12b and to C. Tanasoiu for 6.20 and 6.21. A great debt is also owed to past and present associates such as D. A. McIntyre and P. V. Cooper, who were responsible for working out much of the magnetostatics in Chapters 5 and 6, and thus also to the Ministry of Aviation Supply, the Post Office Research Department, International Computers Limited and the Mullard Research Laboratories, all of whom have provided invaluable support: the latter also provided the crystals used for Figures 6.20 and 6.21.

D.J.Craik
The University of Nottingham

1

Magnetostatic Principles

1.1 CHARGES AT REST: ELECTROSTATICS

Electrostatics is based on the experimental (Coulomb) law governing the forces between electrically charged particles or test bodies. The electron itself would constitute a natural unit of charge but is, in fact, assigned a charge of $4 \cdot 77 \times 10^{-10}$ e.s.u. The law for the force between charges q_1 and q_2, with positions given by the vectors r_1, r_2 (Figure 1.1) is

$$\mathbf{F}_{12} = -\mathbf{F}_{21} = \frac{kq_1q_2}{|\mathbf{r}_1-\mathbf{r}_2|^2}\mathbf{n}. \tag{1.1}$$

This is the vector form of the inverse square law. Note that the possibly more familiar form

$$F = \frac{q_1q_2}{(\text{Separation})^2}$$

carries no information concerning the direction of the force, whereas Equation (1.1), in which $\mathbf{n}$ is a unit vector directed from q_2 to q_1, states that the forces on the two charges are equal in magnitude and directed as shown if q_1 and q_2 are of the same sign. Figure 1.2 should serve to refresh one's memory on the vector conventions.

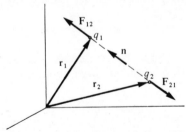

Figure 1.1. Charges q_1 and q_2 located by vectors $\mathbf{r}_1$ and $\mathbf{r}_2$. $\mathbf{F}_{12}$ is the electrostatic force on charge q_1 due to q_2 and vice versa. $\mathbf{n}$ is a vector of unit magnitude directed from q_2 to q_1.

Equation (1.1) can be used to define the unit of charge on the c.g.s. system. In a vacuum $k = 1$ and if two identical charges 1 cm apart give rise to forces of 1 dyne, they are said to have the value of 1 e.s.u.[1]

---

[1] In the rationalized m.k.s. system $k = (4\pi\epsilon_0)^{-1} = 10^{-7}c^2$, and if two identical charges 1 m apart give rise to a force of $10^7 c^{-2}$ newtons, they are assigned the value of 1 m.k.s. unit of charge or 1 coulomb.

The concept of electric field also arises from this equation, for the force on q_1 may be considered to be caused by a field $\mathbf{E}$ at P_1, such that

$$\mathbf{F} = q_1\mathbf{E}. \tag{1.2}$$

Clearly the field $\mathbf{E}$ represents the effect of the charge q_2 at a distance and for consistency with Equation (1.1) $\mathbf{E}$ must be given by

$$\mathbf{E} = \frac{kq_2}{|\mathbf{r}_1 - \mathbf{r}_2|^2}\mathbf{n}, \tag{1.3}$$

which is the inverse square law for the electric field.

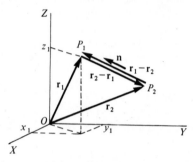

Figure 1.2. Any point such as P_1 is specified more briefly by the vector $\mathbf{r}_1$ than by the coordinates (x,y,z). The designation of the vector P_1P_2 as $(\mathbf{r}_2 - \mathbf{r}_1)$ is verified by noting that a displacement $O \to P_1 \to P_2$ is equivalent to OP_2, as represented by $\mathbf{r}_1 + (\mathbf{r}_2 - \mathbf{r}_1) = \mathbf{r}_2$. The (scalar) distance between P_1 and P_2 is $|\mathbf{r}_2 - \mathbf{r}_1| = |\mathbf{r}_1 - \mathbf{r}_2|$, the vertical lines indicating 'the magnitude of', and the unit vector shown is $(\mathbf{r}_1 - \mathbf{r}_2)/|\mathbf{r}_1 - \mathbf{r}_2| : |\mathbf{n}| = 1$. Note that $|\mathbf{r}_1 - \mathbf{r}_2| \neq |\mathbf{r}_1| - |\mathbf{r}_2|$.

This field can in turn be connected with an electric potential $\phi(x, y, z)$ or $\phi(\mathbf{r})$, a scalar function of position, by the component equations:

$$E_x = -\frac{\partial\phi}{\partial x} \qquad E_y = -\frac{\partial\phi}{\partial y} \qquad E_z = -\frac{\partial\phi}{\partial z}. \tag{1.4}$$

(The field components may be written as scalars, since their directions are given by implication.) These three equations are all included in the single equation

$$\mathbf{E} = -\mathrm{grad}\,\phi = -\left(\mathbf{x}\frac{\partial}{\partial x} + \mathbf{y}\frac{\partial}{\partial y} + \mathbf{z}\frac{\partial}{\partial z}\right)\phi, \tag{1.5}$$

where $\mathbf{x}$, $\mathbf{y}$ and $\mathbf{z}$ are unit vectors along OX, OY and OZ and thus specify the components. Since ϕ is a scalar it is given more simply than $\mathbf{E}$, by

$$\phi = kq/r \tag{1.6}$$

for the potential at a distance r from a point charge q. Obviously the field from a distribution of charges may be obtained by a vector summation (or by integration for a continuous distribution of charge) based on Equation (1.3), or it may be obtained by finding the total potential using

Equation (1.6) and then applying Equation (1.5). These latter summations or integrations of scalar quantities are usually the easier, and this partially explains the great usefulness of potential theory. There is also an important relation with the energy of charges. In view of the forces discussed it is necessary to do work to alter the distance between two charges (or conversely work can be obtained by the spontaneous motion of the charges), and by equating the work done to the energy change of the system the charges can be assigned a mutual potential energy, which becomes zero for infinite separation. Using $dw = \mathbf{F}\,dl = Eq\,dl$, the work done in bringing a charge q from infinity to a point where the field is $\mathbf{E}$ is

$$\int dw = W = q \int_0^E \mathbf{E}\,dl, \qquad (1.7)$$

and in view of Equation (1.5) this may be written

$$W = q\phi, \qquad (1.8)$$

i.e., the energy of a charge at a point where the potential is ϕ is given by the magnitude of the charge $\times$ potential. It is not strictly necessary to refer to the origin of the potential, although it is easier to envisage the interactions between two point charges.

Two point charges, $+q$ and $-q$, of equal magnitude and opposite sign, and with a fixed spacing s, constitute a dipole. From Figure 1.3 it is seen that the turning moment is of magnitude $2(Eqs/2)$, or (qs) in unit field when $\theta = \pi/2$. Thus (qs) is known as the electric dipole moment. For general values of θ the moment is $Eqs\sin\theta$.

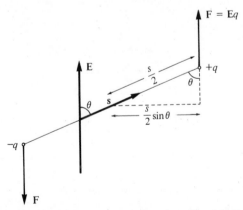

Figure 1.3. A macroscopic electric dipole, of magnitude qs, vector form $\mathbf{d} = s qs$ where s has unit magnitude. Note the direction of the forces with respect to that of the uniform field $\mathbf{E}$, and the convention for the direction of s and thus of the dipole.

Since the vector product of two vectors $\mathbf{A}$ and $\mathbf{B}$, which lie at an angle θ to each other, is defined by

$$\mathbf{A} \times \mathbf{B} = AB\sin\theta,$$

we can make use of this, on indicating the directional nature of the dipole by the expression $sqs = \mathbf{d}$ (see Figure 1.3) and state that the torque on the dipole is given by

$$\mathbf{T} = \mathbf{E} \times \mathbf{d}. \tag{1.9}$$

Figure 1.4, with its caption, gives further information on the vector product.

There is no net translational force on a dipole in a uniform field and thus no potential energy of position, but since work may be required to change the orientation with respect to the field direction there is clearly a potential energy dependent on θ. This is given by the integral of the torque with respect to θ i.e.,

$$W = -Ed\cos\theta \tag{1.10}$$

$$= -\mathbf{E} \cdot \mathbf{d}.$$

where $\mathbf{E} \cdot \mathbf{d}$ is the scalar product of the two vectors with the value indicated (and, of course, no orientation).

Finally, in this section, it should be stressed that the above refers to a 'macroscopic dipole', but a point dipole may be defined as one which has a definite value when $s \to 0$ and $q \to \infty$. In practice, we may consider a point dipole to be one which has negligible extent as compared with the distance from which it is observed.

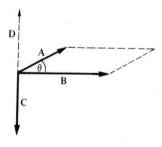

Figure 1.4. The vector product of two vectors $\mathbf{A}$ and $\mathbf{B}$, i.e. $\mathbf{A} \times \mathbf{B}$, is conventionally represented by a vector $\mathbf{C}$ which is normal to the plane containing $\mathbf{A}$ and $\mathbf{B}$ and has magnitude $C = |\mathbf{C}| = AB\sin\theta = $ area of parallelogram shown. Note that $\mathbf{B} \times \mathbf{A} = \mathbf{D} = -\mathbf{C}$. Taking the unit vectors $\mathbf{x}, \mathbf{y}, \mathbf{z}$, directed along orthogonal axes there are relations such as $\mathbf{x} \times \mathbf{x} = \mathbf{y} \times \mathbf{y} = \mathbf{z} \times \mathbf{z} = 0$, $\mathbf{x} \times \mathbf{y} = \mathbf{z}$ etc. and, after expressing $\mathbf{A}$ and $\mathbf{B}$ in cartesian components, it can be shown that

$$\mathbf{A} \times \mathbf{B} = (A_y B_z - A_z B_y)\mathbf{x} + (A_z B_x - A_x B_z)\mathbf{y} + (A_x B_y - A_y B_x)\mathbf{z}$$

or, in the form of a determinant:

$$\mathbf{A} \times \mathbf{B} = \begin{vmatrix} \mathbf{x} & \mathbf{y} & \mathbf{z} \\ A_x & A_y & A_z \\ B_x & B_y & B_z \end{vmatrix}$$

1 .2. CHARGES IN MOTION: MAGNETOSTATICS

1.2.1 Magnetostatic forces

Magnetostatics is based on the observation that when two charges are in motion there exists between them a force which is connected specifically with their motion and is additional to the electrostatic forces. This magnetostatic force obeys the law, as shown by the experiments of Lorentz:

$$\mathbf{f}_{12} = k\left(\frac{q_1 q_2}{|\mathbf{r}_1 - \mathbf{r}_2|^2}\right)\mathbf{v}_1 \times (\mathbf{v}_2 \times \mathbf{n}). \tag{1.11}$$

In this $\mathbf{n}$ is again a unit vector along the line joining the two charges at the instant considered, $\mathbf{v}_1$ is the velocity of q_1 and $\mathbf{v}_2$ that of q_2, and $\mathbf{r}_1, \mathbf{r}_2$ are the position vectors of q_1 and q_2 (Figure 1.5). The constant k is discussed below.

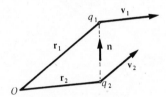

Figure 1.5. Two charges in motion with velocities $\mathbf{v}_1$ and $\mathbf{v}_2$ (not approaching the velocity of light). It is assumed that the motion of each is uniform so that they do not radiate energy.
$\mathbf{n} = (\mathbf{r}_1 - \mathbf{r}_2)/|\mathbf{r}_1 - \mathbf{r}_2|$.

Equation (1.11) is considerably more complex than Equation (1.1). The force is not parallel to the line joining the charges and so, generally

$$\mathbf{f}_{21} \neq -\mathbf{f}_{12}, \tag{1.12}$$

where $\mathbf{f}_{12}$ is the force on charge 1 due to charge 2 and $\mathbf{f}_{21}$ the force on charge 2 due to charge 1. Thus the 'action' and 'reaction' are not usually equal and opposite and Newton's third law of motion does not apply directly. A special case in which $\mathbf{f}_{21} = -\mathbf{f}_{12}$ is illustrated by Figure 1.6: the velocity vectors being parallel.

1.2.2 Magnetic induction, B

The force on the moving charge q_1 can be considered to be caused by a magnetic induction at the point $\mathbf{r}_1$, i.e., $\mathbf{B}(\mathbf{r}_1)$ such that

$$\mathbf{f}_1(\mathbf{r}_1) = \frac{q_1}{c}\mathbf{v}_1 \times \mathbf{B}(\mathbf{r}_1). \tag{1.13}$$

which, with Equation (1.11), gives the definition of the induction as

$$\mathbf{B}(\mathbf{r}_1) = \frac{q_2}{c}\frac{(\mathbf{r}_2 - \mathbf{r}_1) \times \mathbf{v}_2}{|\mathbf{r}_2 - \mathbf{r}_1|^3}. \tag{1.14}$$

The constant c, the velocity of light, by which the charge is divided in Equations (1.13) and (1.14), corresponds to the expression of $\mathbf{f}$ and $\mathbf{B}$ in c.g.s. units, i.e. in dynes and gauss, on the assumption that the charges

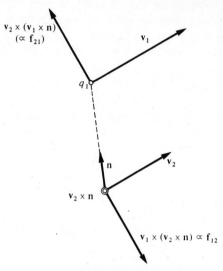

Figure 1.6. The direction of the forces due to the parallel motion of two charges. The vector $\mathbf{v}_2 \times \mathbf{n}$ is directed along the normal to the plane of the diagram, towards the observer.

are in electrostatic c.g.s. units. In order to be consistent the constant k in Equation (1.11) must become $1/c^2$. This is done so that $\mathbf{E}$ and $\mathbf{B}$ should have the same dimensions, as is seen to be the case by comparing Equations (1.2) and (1.13), but it also means that the charge must be taken to have different units and dimensions according to

$$(q:\text{e.m.u.}) = \frac{1}{c}(q:\text{e.s.u.}).$$

In rationalised m.k.s. units q always has the same units but $\mathbf{E}$ and $\mathbf{B}$ have different dimensions: the constant k is expressed as $(1/4\pi)\mu_0$ and given the value 10^{-7} henries m^{-1}. The m.k.s. system has gained wide recognition in engineering, and for problems involving macroscopic charges and currents. However, it is a less natural system for atomic and nuclear studies and practically the entire research literature on magnetic materials continues to be in c.g.s. units; for this reason they are used throughout the current work (see conversion table in Appendix).

If we replace the (charge × velocity) part of Equation (1.14) by a filamentary current element, $i\,\mathrm{d}\mathbf{l}$ as shown in Figure 1.7a, one obtains the contribution to $\mathbf{B}$ (i.e. d$\mathbf{B}$) of the small element as

$$\mathrm{d}\mathbf{B} = \frac{i\,\mathrm{d}\mathbf{l} \times (\mathbf{r}_2 - \mathbf{r}_1)}{|\mathbf{r}_2 - \mathbf{r}_1|^3}. \tag{1.15}$$

with i in e.m.u. $= (i \text{ e.s.u.})/c = (i \text{ amperes})/10$. A little consideration shows that the direction of **B** can be represented as in Figure 1.7b, and if the lines are drawn with a spacing that is directly proportional to the magnitude of **B**, a fair representation of the induction field may be obtained.

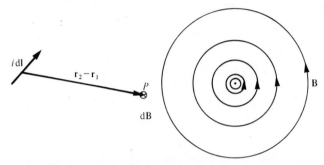

Figure 1.7. (a) The current element $i$dl is situated at the point $\mathbf{r}_1$, and the contribution of this to the induction at the point P ($\mathbf{r}_2$) is directed normal to the diagram, away from the observer. (b) The induction represented by concentric circles around a current which is directed towards the observer.

In many cases it is convenient to consider the current element to lie at the origin, in which case the point P is given by **r** and Equation (1.15) has the alternative forms

$$d\mathbf{B} = \frac{i\,d\mathbf{l} \times \mathbf{r}}{r^3} \qquad (1.16a)$$

and

$$d\mathbf{B} = \frac{i\,d\mathbf{l} \times \bar{\mathbf{r}}}{r^2}, \; |\bar{\mathbf{r}}| = 1. \qquad (1.16b)$$

The integration of these, for conductors of infinite length, demonstrates, for example, that a long straight wire generates fields which fall off as the reciprocal of the distance from the wire and that the lines of field form the closed circles indicated by Figure 1.7b. The most important calculation, in the present context, refers to a circular current loop and is illustrated by Figure 1.8. Although the vector equation defines the

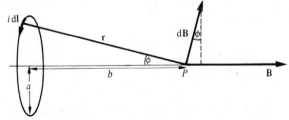

Figure 1.8. The contribution d**B** is normal to **r** and to $i$dl, which can be considered as a linear vector in the limit. By symmetry all the components of d**B** normal to the axis of the coil sum to zero, and **B** is directed as shown.

direction of **B**, it is clear from the caption that the magnitude of **B** is given by the simple integration of the components of the contributions d**B** taken parallel to the axis of the coil, i.e., by

$$B = \int \frac{i \sin\phi}{r^2} \, dl = \frac{i \sin\phi}{r^2} \int dl = \frac{i \sin\phi}{r^2} 2\pi a \tag{1.17}$$

and since $r^2 = a^2 + b^2$, $\sin\phi = a/r$,

$$B = \frac{2\pi i a^2}{(a^2 + b^2)^{3/2}} = \frac{2iA}{(a^2 + b^2)^{3/2}}, \qquad (A = \pi a^2).$$

For points on the axis distant from the loop, $b \gg a$, whence

$$B = \frac{2iA}{r^3} \tag{1.18}$$

where A is the area of the loop and r may now be taken as the distance from the loop.

The quantity iA is of basic importance. It is known as the *magnetic dipole moment* (strictly the magnitude of the dipole), for reasons discussed in the next section.

1.3 DIPOLES AND POLES

1.3.1 Magnetic dipole

The electrostatic dipole, which has already been discussed, generates electric fields which can readily be calculated via the potential of the constituent charges. Referring to Figure 1.9a, the potential at P is

$$\phi = -\frac{q}{r_1} + \frac{q}{r_2} = q \times (\text{change in } 1/r \text{ in going from } A \text{ to } B)$$

that is

$$\phi = qs \frac{\partial}{\partial s}\left(\frac{1}{r}\right).$$

In the limit when $s \to 0$ and $qs \to d$, i.e. for a point dipole, the potential is given by the vector equation

$$\phi = \mathbf{d} \cdot \mathrm{grad}_A (r^{-1}), \tag{1.19}$$

where grad_A means that the gradient must be evaluated with respect to the coordinates of A. Representing the relative position of P with respect to A by the vector $\mathbf{r}$ (which may be considered as the position of the dipole when $s \to 0$), $\mathrm{grad}_A (r^{-1})$ becomes $\mathbf{r} r^{-3}$ and Equation (1.19) becomes

$$\phi = \frac{\mathbf{d} \cdot \mathbf{r}}{r^3} = \frac{d \cos\theta}{r^2}. \tag{1.20}$$

The field can then be found from $E = -\text{grad}\,\phi$ but it is easier to find the radial and transverse resolutes as

$$E_r = -\frac{\partial\phi}{\partial r} = \frac{2d\cos\theta}{r^3},$$

$$E_\theta = -\frac{1}{r}\frac{\partial\phi}{\partial\theta} = \frac{d\sin\theta}{r^3}. \qquad (1.21)$$

When $\theta = 0$, for points on the axis, we have simply

$$E = \frac{2d}{r^3}. \qquad (1.22)$$

Now it can be seen that Equations (1.18) and (1.22) are entirely analogous if (iA) in Equation (1.18) is regarded as a magnetic dipole, having similar properties to those of the electric dipole d, although the basic approach is so different in the two cases. Furthermore, in view of the coincidence demonstrated, it should be easy to accept that if the B field distribution is calculated generally, for points distant from the loop in comparison with its diameter, this has precisely the same form as that indicated by Figure 1.9b and Equation (1.21), i.e. we may write

$$B_r = \frac{2\mu\cos\theta}{r^3},$$

$$B_\theta = \frac{\mu\sin\theta}{r^3}. \qquad (\mu = iA). \qquad (1.23)$$

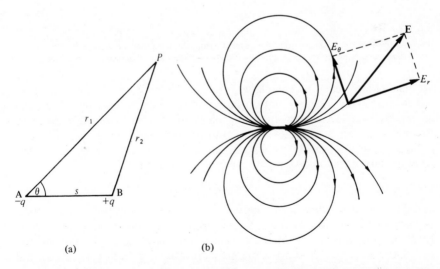

(a)　　　　　　　　(b)

Figure 1.9. (a) Macroscopic electrostatic dipole, for the calculation of the potential at the point P. (b) The field distribution due to a point (electrostatic or magnetostatic) dipole, in terms of its radial and transverse components E_θ and E_r, or B_θ and B_r.

There is yet another correlation between electrostatic dipoles and current loops. By recourse to the macroscopic case it was shown that the electrostatic dipole was subjected to torques in the presence of uniform fields. The forces on conductors, which are in the form of a rectangle and are oriented in a magnetic induction field as shown in Figure 1.10,

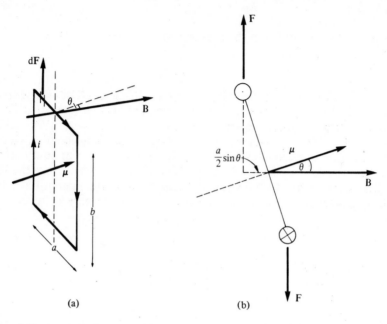

(a)   (b)

Figure 1.10. A rectangular current loop shown in perspective (a) and plan (b). The forces on the top and the bottom conductors cancel, while the forces on the sides give a turning moment (no net translational force). μ shows the direction of the equivalent dipole.

are particularly easy to calculate. On any element of the top side there is a force exerted in the direction shown in the Figure, but there is clearly an equal force on a corresponding element of the lower side directed downwards. Since all these contributions are equal and opposite and collinear, there is neither a net force nor a net torque. The forces on the sides are also equal and opposite having the magnitude

$$F = ibB$$

and the direction shown in Figure 1.10b. The important point is that, in view of the nature of the cross product, these forces are again normal to B but are not in the plane of the loop. Thus there is again no net translational force (which is obviously the case for Figure 1.3 also), but there is a torque given by

$$T = ibB2(a/2)\sin\theta = iAB\sin\theta.$$

A is the area of the loop, and so again iA can be written as μ, the magnetic dipole moment and

$$T = \mu B \sin\theta. \tag{1.24}$$

We now assign a direction to μ, which is simply that required to achieve consistency with the direction of the fields generated by the dipole in analogy with the electrostatic case, and is as shown in Figure 1.10. Since θ is the angle between $\mathbf{B}$ and μ, Equation (1.24) may be given the vector form

$$\mathbf{T} = \mu \times \mathbf{B}. \tag{1.25}$$

The same considerations apply to current loops of any shape, as can be seen by dividing them up into rectangular strips and noting that the effects of the current cancel except for the peripheral sections.

1.3.2 Magnetic pole and potential

The very similar behaviour of magnetic and electric dipoles, with respect to the generation of fields and the existence of torques in applied fields, prompts one interesting question. Why was it not possible to make an exactly analogous approach in the two cases, using magnetic equivalents to the point charges? This could have been done in principle, but it would have appeared rather forced because it would have necessitated commencing with entities, the magnetic monopoles, which do not exist in nature. Even on the scale of size of the atom the electron may be treated as a point charge for many purposes, and an electron rotating about an atomic nucleus constitutes a fair approximation to a point magnetic dipole. A 'spinning' electron is even closer to an effective point dipole, but there is no conceivable way in which the magnetic dipole can be resolved into separate positive and negative monopoles. However, it is possible to define hypothetical poles, always associated with real dipoles, by analogy with the electrostatic case. These poles are then a considerable aid to computation, although the limitations imposed by their hypothetical nature must always be borne in mind. For example, it is immediately obvious that there is no possibility of any mobility of the poles across a surface or throughout a material: their rigid association with the real dipoles is essential.

One property of dipoles is that they are additive (vectorially) in the sense that, if two identical current loops are superimposed, they clearly give fields which are everywhere double that given by one current loop; alternatively, referring to Equation (1.20) the potential given by two coincident dipoles is

$$\phi = \frac{\mu_1 \cdot \mathbf{r}}{r^3} + \frac{\mu_2 \cdot \mathbf{r}}{r^3} = \frac{(\mu_1 + \mu_2) \cdot \mathbf{r}}{r^3}$$

equivalent to that of a dipole $(\mu_1 + \mu_2)$. Further, if we stack dipoles together, as shown in Figure 1.11, each will give rise to its own field and potential (since magnetic fields, like electric fields, superpose without interference) and at distances considerably greater than the dimensions of the array the effects will be those of a single dipole of value $nm\mu$.

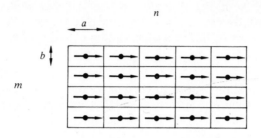

Figure 1.11. An array of magnetic dipoles, $n \times m$, representing a magnetized material with an induced magnetic moment $nm\mu$.

From this point of view the dipoles add even when not coincident, and the dipole moment of the array may be defined by

$$\mathbf{M} = \sum \mu. \tag{1.26}$$

We further define the intensity of magnetization (for a three-dimensional array) as the dipole moment per unit volume:

$$\mathbf{I} = \mathbf{M}/V = \sum \mu/V. \tag{1.27}$$

The dipoles need not all have the same magnitude and orientation for this to apply; for example reversing each alternate dipole in Figure 1.11 would give $\mathbf{M} = 0$ and $\mathbf{I} = 0$, since they are given by vector sums.

Assuming, for the moment, a regular array (i.e., uniform $\mathbf{I}$) with spacing $1/n$, the total dipole moment must be subjected to a torque in a uniform field of

$$\mathbf{T} = \mathbf{M} \times \mathbf{B} = n^3 \mu V \times \mathbf{B}, \tag{1.28}$$

since the number of dipoles per unit volume is n^3. The number per unit cross-sectional area is n^2, and if the array has a cross-sectional area A and length l, $V = Al$, and

$$\mathbf{T} = n^2 Al\mu \times \mathbf{B} = n^2 \mu Al \times \mathbf{B}. \tag{1.29}$$

Consequently the total dipole moment may be considered either as the magnetization × volume, or as that of a pair of monopoles of magnitude $n^2 \mu A$ and separation l. For this reason, the ends of the array may be considered to have a pole density, σ, of magnitude I per unit area. If the surface lies at an angle θ to the direction of the dipoles, and thus to $\mathbf{I}$, then

$$\sigma = \pm \mathbf{I} \cos\theta = \pm \mathbf{I} \cdot \mathbf{n},$$

where $\mathbf{n}$ is the unit vector normal to the surface. The sign convention adopted is that σ is positive if $\mathbf{I}$ is parallel to the outward normal to the surface, and negative if $\mathbf{I}$ is antiparallel to the outward normal.

This pole density makes an important contribution to the potentials and induction arising from dipole arrays. Since any element of the surface, ds, has an associated pole of strength $\mathbf{I} \cdot \mathbf{n}ds$, it gives rise to a potential at a point distant r, of

$$d\phi = \frac{\mathbf{I} \cdot \mathbf{n}}{r} ds$$

or

$$\phi = \int \frac{\mathbf{I} \cdot \mathbf{n}}{r} ds \qquad (1.30)$$

from the surface as a whole. When the magnetization is non-uniform, it gives rise to a volume distribution of poles of volume density $\mathrm{div}\mathbf{I}$ where

$$\mathrm{div}\mathbf{I} = \frac{\partial I_x}{\partial x} + \frac{\partial I_y}{\partial y} + \frac{\partial I_z}{\partial z}. \qquad (1.31)$$

In this case, the potential produced by the (non-uniform) array of dipoles is the sum of two terms i.e.

$$\phi = \int \frac{\mathbf{I} \cdot \mathbf{n}}{r} ds + \int \frac{\mathrm{div}\mathbf{I}}{r} dV. \qquad (1.32)$$

Clearly, when $\mathbf{I}$ is uniform the rate of change of any of its components is zero, $\mathrm{div}\mathbf{I} = 0$, and the only contribution is from the surface poles.

Since it is shown in the following section that a magnetic field $\mathbf{H}$ can be related to ϕ by $\mathbf{H} = -\mathrm{grad}\,\phi$, and also that the divergence $\mathrm{div}\mathbf{H}$ is zero, it follows that $\mathrm{div}\,\mathrm{grad}\,\phi = 0$, that is, in the absence of magnetic material

$$\mathrm{div}\,\mathrm{grad}\,\phi = \left(\frac{\partial^2}{\partial x^2} + \frac{\partial^2}{\partial y^2} + \frac{\partial^2}{\partial z^2} \right)\phi = 0.$$

This is Laplace's equation, which is more briefly written as

$$\nabla^2 \phi = 0,$$

and it follows that many of the theorems and computations developed in the study of electrostatics, based on the solutions of Laplace's equation, apply also to the magnetic cases. The great importance of such solutions, where the appropriate boundary conditions are used, in the calculation of magnetostatic energies for complicated magnetic structures will be appreciated by reference to Chapter 6. The basic problem is the integration of the product of potential and surface charge, $\phi \times \sigma$ or $\phi \times \mathbf{I} \cdot ds$, when we have obtained σ in the form of a series and ϕ as a solution of Laplace's equation appropriate to the particular structure.

1.4 MAGNETIC FIELD, H, AND MAGNETIC INDUCTION, B

It has been seen that magnetic induction is associated with macroscopic electric currents, with magnetic dipoles which are equivalent to microscopic current loops (e.g. orbiting or spinning electrons), and with surface pole distributions. Particularly where dipoles and poles are concerned, one may wonder why the analogy with electrostatics is not maintained to the extent of describing the distant effects in terms of a magnetic field, rather than induction.

There are, in fact, two alternative approaches to the subject. The first, as in the m.k.s. system, deals solely with the quantity magnetic induction, in which case the difference between the principles applying to the induction in the presence or absence of magnetized material is rather difficult to clarify. The second, which is used here, makes a somewhat arbitrary but quite definite distinction between magnetic field, H, and magnetic induction.

The term magnetic field is used to describe the magnetic effects of currents, dipoles, and poles in regions of space which contain no magnetic material (i.e. dipole arrays). This applies to vacua and, for most purposes, effectively to weakly magnetic gases. With this restriction, the definition and equations given for B can be rewritten for H, e.g. Equation (1.18) becomes $H = 2iA/r^3$ and the dipole field components (Equation 1.23) are $H_r = 2\mu\cos\theta/r^3$, and $H_\theta = \mu\sin\theta/r^3$. Furthermore, the field H can be related to the magnetic potential, ϕ, by

$$H = -\text{grad}\,\phi, \qquad (1.33)$$

where ϕ is derived from dipole arrays with volume divergence or surface discontinuities, as by Equation (1.32).

The same definition and methods of computation still apply in the presence of uniformly magnetized material of infinite extent ($\text{div}\,I = 0$ and $I \cdot n = 0$), but now the effects of the macroscopic currents, for example, are no longer fully described by the field alone. Suppose that a pair of coils or a large solenoid is used to generate a uniform magnetic field and a long rod of a (para-) magnetic material, with negligible end effects, is placed in the field. This material consists of an array of dipoles, associated with the atoms and assumed to be in a regular array as in a crystal, and these are initially oriented at random. The field exerts a torque on each dipole equal to $\mu \times H$ and thus causes a certain mean orientation which is opposed by thermal agitation (see section 2.2). All components normal to the field direction cancel and leave only the components in the field direction, say m where $m = \langle \mu \cdot H \rangle / H$ i.e. the field induces a certain level of magnetization $I = mn$, which is parallel to the field direction (with n dipoles/unit volume).

It must now be recalled that the dipoles consist of microscopic current loops, which themselves generate fields. The magnitude of this contribution may be derived by a simple demonstration illustrated by

Figure 1.12. For the single line of dipoles the magnetization, or dipole density per unit volume, is $I = Nm/A = N(iA)/A = Ni$. But the line is also equivalent to a solenoid with current density Ni in which the field is $4\pi Ni$ or $4\pi I$, using the above. This can be seen to hold for a three-dimensional array also; the effects of oppositely-directed neighbouring currents cancel leaving only an effective surface current. The useful convention is to refer to the sum of the applied fields together with this contribution from the material itself as the induction i.e.

$$\mathbf{B} = \mathbf{H} + 4\pi\mathbf{I}. \tag{1.34}$$

The value of the convention is that it obviates the necessity to refer continuously to the "effect associated with macroscopic currents etc." and to the "effect associated with magnetized material". It would be necessary to keep making this distinction because, although our choice of what was to be referred to as **H** and what as **B** was originally arbitrary, having made the choice it can be seen that **H** and **B** do have different properties.

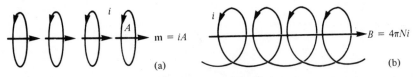

Figure 1.12. (a) A linear array of dipoles of magnitude $m = iA$; N per unit length, i.e. in a volume $A \times 1$. (b) Equivalent solenoid with current density Ni and induction $4\pi Ni$.

It will be shown in the next section that the fields, **H**, arising from a surface with pole density I, have magnitude $2\pi I$ and the directions shown in Figure 1.13 (at points very near to the surface, or anywhere if the surface has unlimited extent). Consequently the surface constitutes a discontinuity so far as **H** is concerned, but it is also clear from the figure that **B** is continuous across the surface. Taking, for convenience, the simple 'pill-box' surface shown in Figure 1.14, the integrals taken over closed surfaces are

$$\int_{c.s.} \mathbf{B} \cdot d\mathbf{s} = 0 \tag{1.35}$$

and

$$\int_{c.s.} \mathbf{H} \cdot d\mathbf{s} = 4\pi I \tag{1.36}$$

if the chosen surface encloses unit area of the material surface, or more generally $4\pi \times$ (total surface pole enclosed). Thus, using Green's theory, which states that for any such vector quantity

$$\int \mathbf{B} \cdot d\mathbf{s} = \int \operatorname{div} \mathbf{B} \, dv, \tag{1.37}$$

the above is equivalent to

$$\operatorname{div} \mathbf{B} = 0 \text{ everywhere,} \qquad (1.38)$$

whereas

$$\operatorname{div} \mathbf{H} = 0 \qquad (1.39)$$

only in the absence of magnetic material, since then $\mathbf{I} = 0$ and $\mathbf{B} = \mathbf{H}$. Otherwise $\operatorname{div} \mathbf{H} = 4\pi\sigma$ and $\Delta^2\phi = -4\pi\sigma$ (Poisson's equation).

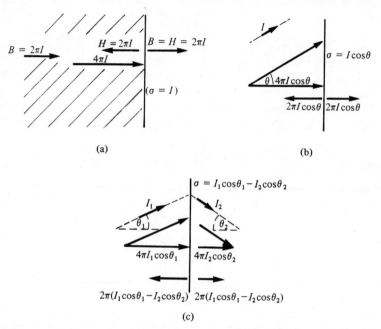

(a)

(b)

(c)

Figure 1.13. Simple geometric illustrations of the continuity of the normal component of the induction, $\mathbf{B} = \mathbf{H} + 4\pi\mathbf{I}$ across the surface of a magnetic material (a, b) or across an interface between two materials with differing directions and magnitudes of the magnetization, with $B = 2\pi(I_1\cos\theta_1 + I_2\cos\theta_2)$ across the interface.

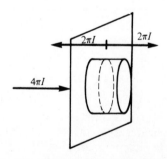

Figure 1.14. Simple surfaces across which the integrals of **B** and **H** can be immediately inferred.

An interesting property of fields and potentials can be demonstrated by referring back to Figure 1.7. The product $\mathbf{H} \cdot \mathbf{dl} = H\,dl\cos\theta$ gives the component of the field in the direction of the arbitrary length or line vector element $\mathbf{dl}$. Taking the line integral, $\int\mathbf{H} \cdot \mathbf{dl}$, around any circle centred on, and in a plane normal to, the current, is particularly simple since $\mathbf{H}$ is constant in magnitude ($i/2\pi r$) and is always parallel to the line ($\cos\theta = 1$):

$$\int\mathbf{H} \cdot \mathbf{dl} = \frac{i}{2\pi r}\int dl = i. \tag{1.40}$$

Any path surrounding a conductor can be shown to give the same result, by dividing the path into sections as shown in Figure 1.15, and Ampere's circuital law is of universal application. In particular

$$\int\mathbf{H} \cdot \mathbf{dl} = 0 \tag{1.41}$$

in the absence of macroscopic currents. This applies whether magnetic material is present or not (on the assumption that one cannot devise a path which threads a microscopic current loop).

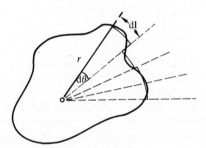

Figure 1.15. Any arbitrary path can be approximated by a series of radial segments, in the direction of which $H = 0$, and circular sections along which $\int\mathbf{H} \cdot \mathbf{dl}$ has constant value since $H \propto 1/r$ and the length $dl \propto r$: thus $\int\mathbf{H} \cdot \mathbf{dl}$ always has the same value as that for a circular path, which is equal to the current enclosed.

The relation $\mathbf{H} = -\mathrm{grad}\,\phi$ may be considered to lead to the definition of the magnetic potential:

$$\phi = \int\mathbf{H} \cdot \mathbf{dl} \tag{1.42}$$

since $\mathbf{H}$ can be independently defined in terms of the force between currents. According to the preceding equations this potential will only be single-valued in the absence of macroscopic currents. Consequently, the definition of ϕ is only meaningful in this context, i.e. when it is associated with dipoles or pole distributions.

1.5 MAGNETIC ENERGY

The orientational dependence of the energy of a magnetic dipole in a uniform field, H_a, can be derived directly from the torque (section 3.1) as

$$W = -\mu \cdot H_a, \tag{1.43}$$

the negative sign ensuring that the energy is minimal (equal to $-\mu H$) when μ and H are parallel. Extending this to an array of dipoles which are represented both in direction and density by I (which may vary in magnitude or direction from place to place)

$$W = -\int H_a \cdot I \, dV. \tag{1.44}$$

For uniform magnetization this is simply $H \cdot I$ for unit volume.

An array of oriented dipoles, or magnetized body, has a so-called self energy even in the absence of an applied field. For a uniform array this can be considered to arise in two different, but equivalent, ways. First, each element of surface ds can be treated as a surface pole of value $I \cdot ds$, which is situated in the potential created by the poles on the remainder of the surface. Then the total energy will be given by integrating the product of the elementary poles and potentials as

$$W = \tfrac{1}{2} \int \phi I \cdot ds. \tag{1.45}$$

The factor of one half arises because otherwise each surface element would have been considered twice; once as a source of potential and again as an elementary pole in the potential.

From a different point of view, the surface poles are noted to produce a field within the material of value H_i. The energy of interaction of the magnetization with this field is, by analogy with Equation (1.44) and again introducing the statistical factor of one half, given by

$$W = \tfrac{1}{2} \int H_i \cdot I \, dV. \tag{1.46}$$

This expression was also derived from first principles by Stoner [1] as representing the work necessary to assemble the array of dipoles.

It is possible to derive yet another expression for the energy, in terms of the fields arising from the surfaces, as

$$W = \frac{1}{8\pi} \int H_i^2 dV, \tag{1.47}$$

where the integration is over all space and not just within the array. (Use $B = H_i + 4\pi I$ and Equation (1.46), and show that $\int B \cdot H_i \, dV = 0$: $B \cdot H_i = -B \cdot \mathrm{grad}\,\phi$). By Green's theorem $\mathrm{grad}\,\phi = -\mathrm{div}(\phi B) + \phi\,\mathrm{div}\,B$; $\mathrm{div}\,B = 0$ and $-\int \mathrm{div}(\phi B)dV = \int \phi B \cdot ds$, but at large distances $\phi \to 0$ and $B \to 0$.)

1.6 FIELDS FROM MAGNETIZED MATERIAL

Any magnetic specimen, either with induced or permanent magnetization, can be considered to give rise to fields with the distribution characteristic of a dipole at points which are far distant compared with the dimensions of the specimen. However, at points close to a rod-shaped specimen the fields are better represented by the superimposition of two monopole fields. Dipole fields were described in section 1.3, and require no further discussion.

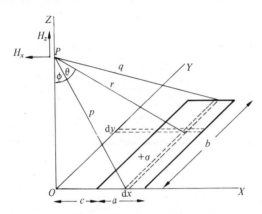

Figure 1.16. Illustration for the calculation of the field (i.e. components parallel to OX and OZ) for a plane rectangular sheet of uniform pole density $+\sigma$ (i.e. 'north poles').

For a uniformly magnetized specimen ($\mathrm{div}\,\mathbf{I} = 0$) the external or stray fields and the internal, or demagnetizing, fields can be calculated from the effective pole density, $\sigma = \mathbf{I} \cdot \mathbf{n}$, at the specimen surfaces. A useful example is the calculation of the field components from a rectangular sheet of uniform pole density, with sides a and b and distant c along OX (Figure 1.16). The field at P due to the elementary pole $\sigma\,dx\,dy$ at (x,y) is $\sigma\,dx\,dy/r^2$ along r. With $y = p\tan\theta$, then $dH_r = \sigma\,dx\,d\theta/p$ and the component along p is $dH_p = \sigma\,dx\cos\theta\,d\theta/p$. Integrating with respect to θ gives the field due to the strip as $\sigma b\,dx/pq$ with a component parallel to OX as

$$dH_x = \sigma bx\,dx/p^2 q.$$

Integrating with respect to x from c to $(c+a)$ gives, for the whole sheet

$$H_x = -\sigma\log\left\{\frac{[(c+a)^2+z^2]^{1/2}}{[(c+a)^2+z^2+b^2]^{1/2}+b} \cdot \frac{(c^2+z^2+b^2)^{1/2}+b}{(c^2+z^2)^{1/2}}\right\}, \quad (1.48)$$

the negative sign indicating that the field is antiparallel to the positive OX direction for positive charges. The resolved field along OZ due to the strip is

$$\frac{\sigma bz\,dx}{p^2 q} = \frac{\sigma bz\,dx}{(x^2+z^2)(x^2+z^2+b^2)^{1/2}}.$$

Substituting $x = z\tan\phi$ and integrating with respect to ϕ gives

$$H_z = \sigma\sin^{-1}\left\{\frac{(c+a)b}{[(c+a)^2+z^2]^{\frac{1}{2}}(b^2+z^2)^{\frac{1}{2}}}\right\} - \sigma\sin^{-1}\left\{\frac{bc}{[(c^2+z^2)(b^2+z^2)]^{\frac{1}{2}}}\right\}$$

$$(1.49)$$

the limits of ϕ being $\sin^{-1}\left\{\frac{c}{(z^2+c^2)}\right\}^{\frac{1}{2}}$ and $\sin^{-1}\left\{\frac{(c+a)}{[z^2+(c+a)^2]^{\frac{1}{2}}}\right\}$. The first term is always the larger and when σ is positive (north poles) the direction is parallel to positive OZ.

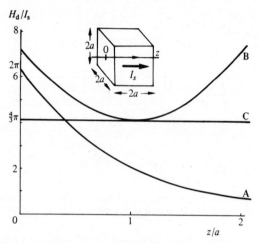

Figure 1.17. Reduced internal demagnetizing fields (H_d/I_s) calculated for points along the central axis of a uniformly magnetized cube, i.e. the axis OZ as shown inset. A: field derived from one surface normal to $\mathbf{I}_s$, $H_d/I_s = 4\sin^{-1}[a^2/(a^2+z^2)]$; B: sum of fields from both such surfaces, and C: demagnetizing fields for a sphere, which is uniform and coincides with that at the centre of the cube.

Naturally these expressions reduce drastically in conditions of high symmetry. Consider the cube shown inset in Figure 1.17, with uniform magnetization $\mathbf{I}_s \parallel OZ$. In this case $\sigma = \mathbf{I}_s \cdot \mathbf{n} = I_s$ and Equation (1.49) gives, for four squares of charge with side a,

$$H_z = 4I_s\sin^{-1}\left(\frac{a^2}{a^2+z^2}\right). \qquad (1.50)$$

Since this is directed opposite to the magnetization, it may be called a demagnetizing field and is plotted as H_d/I_s in the figure (curve A). By symmetry this is the only field component along the central axis. The total demagnetizing field is the sum of the fields from the two surfaces (curve B). It is to be noted that the limiting value of H_z is $2\pi I_s$, and for an infinitely thin sheet the internal, demagnetizing, field is uniform and equal to $4\pi I_s$ while the external field is zero.

Note also that H_x, in Equation (1.48), is unlimited and shows a logarithmic approach to infinity very close to the edge of a sheet of charge or surface of a magnetized specimen.

1.6.1 Demagnetizing fields and energies in ellipsoids of revolution

In principle, the self energy or demagnetizing energy of the uniformly magnetized cube could be found as the integral of $\frac{1}{2}\mathbf{H_d} \cdot \mathbf{I_s}$ or $\frac{1}{2}H_z I_s$ over the whole specimen. This would obviously be very difficult, since H_z varies in a different way along different axes parallel to OZ, and also for a real material the field components normal to OZ are expected to disturb the postulated uniformity of the magnetization and thus to affect the energy.

There is, however, just one general shape for which the demagnetizing fields can be shown to be uniform in direction and magnitude, so long as the material is uniformly magnetized. (It follows that unless the anisotropy forces aligning the magnetization are infinitely strong, it is only then possible to achieve completely uniform magnetization in such bodies.) The shape is an ellipsoid, any particular case being specified by the three major axes a, b and c. Usually discussion can be confined to oblate or prolate spheroids, which are generated by rotating an ellipse about a principle axis and are specified by a polar axis (or semi-axis) a, an equatorial semi-axis b, or by their ratio $q = a/b$. For such bodies the demagnetizing fields are given by calculable demagnetizing coefficients, N_a and N_b for magnetization parallel to the polar or equatorial axes, as $H_d = -N I_s$.

For prolate spheroids (with $q > 1$)

$$N_a = \frac{4\pi}{(q^2-1)}\left\{\frac{q}{(q^2-1)^{1/2}}\log[q+(q^2-1)^{1/2}]-1\right\}, \qquad (1.51)$$

$$N_b = \tfrac{1}{2}(4\pi - N_a) \qquad (1.52)$$

and for oblate spheroids ($q < 1$)

$$N_a = \frac{4\pi}{(1-q^2)}\left[1-\frac{q}{(1-q^2)^{1/2}}\cos^{-1}q\right]. \qquad (1.53)$$

An approximation which is good for values of $q > 2$ is

$$N_b = \frac{2}{q}\left(3-\frac{1}{q}\right). \qquad (1.54)$$

Demagnetizing factors for prolate spheroids are given in Table 1.1.

Values of N_a are plotted in Figure 1.18 as $H_d/4\pi I_s = N_a/4\pi$. Obvious limiting cases are $N_a = 4\pi$ for a very thin sheet ($q \to 0$) and $N_a \to 0$ for very long rods ($q \to \infty$). A very simple value is $N_a = N_b = 4\pi/3$ for a sphere.

Table 1.1. Demagnetizing factor $N_a/4\pi$ for an ellipsoid of revolution polar semi-axis a, equatorial semi-axis b; $q = a/b$. N_b may be calculated from
$$N_a + 2N_b = 4\pi.$$

q	$\dfrac{N_a}{4\pi} \times 10^4$	q	$\dfrac{N_a}{4\pi} \times 10^4$	q	$\dfrac{N_a}{4\pi} \times 10^5$	q	$\dfrac{N_a}{4\pi} \times 10^6$
0·01	9845	1·0	3333	3·5	8965	16	9692
0·02	9694	1·1	3083	4·0	7541	18	8013
0·05	9262	1·2	2861	4·5	6445	20	6749
0·1	8608	1·4	2488	5·0	5582	25	4671
0·2	7505	1·6	2187	5·5	4889	30	3444
0·3	6614	1·8	1941	6·0	4323	·35	2655
0·4	5882	2·0	1736	7·0	3461	40	2116
0·5	5272	2·2	1563	8·0	2842	45	1730
0·6	4758	2·2	1417	9·0	2382	50	1443
0·7	4321	2·4	1291	10·0	2029	60	1053
0·8	3944	2·8	1182	12·0	1530	80	637
0·9	3618	3·0	1087	14·0	1200	100	430

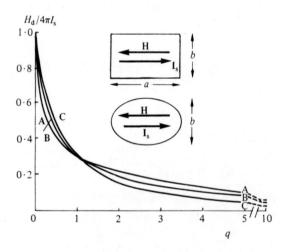

Figure 1.18. Reduced demagnetizing fields *at the centres* of uniformly magnetized prismatic crystals of square cross section (curve A), compared with the demagnetizing fields (at any point, since they are uniform) in the inscribed ellipsoids (curve C). Also shown (curve B) are the 'equivalent demagnetizing fields' for blocks obtained from the calculated demagnetizing energies. All three curves coincide at $q = a/b = 1$, since the demagnetizing energy for a cube is $\frac{1}{2}(4\pi/3)I_s^2$.

Also shown in Figure 1.18 (curve A) are the reduced demagnetizing fields at the centres of uniformly magnetized, rectangular, square cross section, prismatic crystals which are given by

$$\frac{H_d}{4\pi I_s} = \frac{2}{\pi}\sin^{-1}\left(\frac{1}{1+q}\right),$$
(1.55)

where q is the dimensional ratio (see inset). There is moderately good agreement at the centres of the specimens (precise agreement for $q = 1$, as shown also in Figure 1.17). However, it must be stressed, that the fields in the blocks vary in magnitude and direction. It is only for ellipsoidal specimens that the demagnetizing fields are uniform, and the demagnetizing energies are thus given simply by $\frac{1}{2}N_d I$. The simplest such shape to prepare in practice is that with $q = 1$, i.e. a sphere.

1.6.2 Demagnetizing energies for rectangular blocks

Returning to the problem of the rectangular block, and continuing to neglect the likelihood of deviations from uniform magnetization, the demagnetizing energies can be calculated by direct integration, although they cannot be expressed analytically. The mutual potential energy of two surfaces is

$$W = \iint \sigma_2 \phi_2(x_2, y_2)\,dx_2 dy_2,$$
(1.56)

where ϕ_2 is the potential at the point (x_2, y_2) on the surface 2, arising from the charges on the surface 1. Fortunately, for the very useful case illustrated by Figure 1.19, the integrals have been evaluated in terms of the parameters $p = b/a$, $q = c/a$, $r = d/a$ [3].

The mutual energy of two identical rectangles, lying opposite to each other, with uniform charge σ is given as

$$W = 2a^3\sigma^2 f(p,q),$$
(1.57)

and the self energy of a uniformly charged rectangle is

$$W = a^3\sigma^2 f(p, 0).$$
(1.58)

The mutual energy of two similarly oriented rectangles of length a and widths b_i and b_j and separated by a distance d_{ij} as shown in the figure, is

$$W = a^3\sigma^2[f(p_i + p_j + r, q) + f(r,q) - f(p_i + r, q) - f(p_j + r, q)].$$
(1.59)

(This is not needed for the present purpose, but is used in Chapter 5.)

The function f is

$$f(p,q) = (p^2-q^2)\phi[1/(p^2+q^2)^{\frac{1}{2}}]+p(1-q^2)\phi[p/(1+q^2)^{\frac{1}{2}}]$$
$$+pq^2\phi(p/q)+q^2\phi(1/q)+2pq\tan^{-1}[q(1+p^2+q^2)^{\frac{1}{2}}/p]$$
$$-\pi pq-\tfrac{1}{3}(1+p^2-2q^2)(1+p^2+q^2)^{\frac{1}{2}} \tag{1.60}$$
$$+\tfrac{1}{3}(1-2q^2)(1+q^2)^{\frac{1}{2}}+\tfrac{1}{3}(p^2-2q^2)(p^2+q^2)^{\frac{1}{2}}+\tfrac{2}{3}q^3,$$

where

$$\phi(x) = \sinh^{-1}x,$$

$$p = b/a, \quad q = c/a \quad \text{and} \quad r = d/a.$$

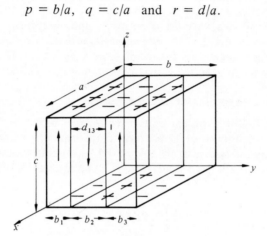

Figure 1.19. A rectangular block of magnetic material with sheets of positive and negative pole density on two surfaces, as produced by subdivision into domains with alternately anti-parallel directions of magnetization. It is assumed that the magnetization is uniform within each domain, and that the boundaries between the domains are of negligible thickness.

For the uniform block it is only necessary to evaluate the two (equal) self energies, and the single interaction energy. All self energies are positive whereas interaction energies are positive when the signs of the charges are the same, and negative when they are opposite, as here.

We can now derive an 'equivalent demagnetizing field' from the calculated energy as

$$H = 2W/I_s.$$

This is plotted as curve B in Figure 1.18: it is not a real field but it turns out that it is very close to the actual field at the centre of the block (curve A) calculated from Equation (1.55), and also close to the uniform demagnetizing field for an ellipsoid with the same axial ratio.

A rather remarkable coincidence has been stressed by Figure 1.17, in which the straight line represents the uniform field in a sphere: just at the centre of a cube the demagnetizing field is identical to that for the sphere.

A further coincidence is shown by Figure 1.18; as noted by Rhodes and Rowlands [3] the demagnetizing energy for a cube is $\frac{1}{2}(4\pi/3)I_s$, as for a sphere, and so all three curves coincide at $q = 1$.

It should be stressed that there are always at least the two equivalent methods for calculating demagnetizing or magnetostatic energies. One way is to integrate the product of the elementary surface pole and the potential, at that point, produced by the remaining pole distribution. This is the method used here, and it is of great value in magnetic analyses generally: a disadvantage is that it hardly ever leads to analytical expressions for the energy, i.e. the computations must be carried out for each individual case but, with access to electronic computers, this is not too onerous. In principle the method could be applied to the calculation of demagnetizing energies of ellipsoids, but since these can be obtained in analytical form the effort would be superfluous.

The second method consists of integrating the scalar product, $\mathbf{H}_d \cdot \mathbf{I}$, of the demagnetizing field and the magnetization. This is the method employed for ellipsoids and, again in principle, it could be applied to rectangular blocks, but it would then require the computation of the entire field distribution which would be extremely difficult, even if the magnetization were assumed to be uniform in direction. In practice such calculations are usually of importance for ferromagnetic materials which have a spontaneous magnetization, the magnitude of which is virtually independent of applied fields: they can be represented by arrays of (atomic) dipoles of identical magnitude and maintained in spontaneous mutual alignment by exchange forces. This produces a net spontaneous magnetization, $\mathbf{I}_s$ equal to the dipole density, which is also aligned as a whole with certain axes of the crystal, i.e. easy directions, by magneto-crystalline anisotropy[2]. However, this anisotropy is always finite and may be quite low, so that in the presence of non-uniform demagnetizing fields $\mathbf{I}_s$ will vary in direction though not in magnitude, as the exchange forces are always much stronger than magnetostatic forces. The real distribution of the magnetization in a rectangular ferromagnetic crystal will thus be as shown in Figure 1.20.b, as compared with Figure 1.20.a, which assumes infinite anisotropy. The complexity of calculations of demagnetizing energy is then apparent, and in practice most detailed magnetostatic analyses have been made for materials in which the rotation of the magnetization can be neglected ($\mathrm{div}\,\mathbf{I} = 0$).

In ellipsoids, however, the demagnetizing fields are uniform in direction and magnitude, and for specimens of this particular shape the magnetization should remain uniform whatever the value of the anisotropy (Figure 1.20.c).

Some of the most complex magnetostatic calculations apply to ferromagnetic crystals, which are subdivided into domains or regions with different directions of magnetization. Figure 1.19 may be taken as an

---

[2] These principles are developed in the succeeding chapters.

example of a simple domain structure. There is then a considerable difference between the methods of calculation appropriate to finite specimens, effectively those containing just a few domains, and specimens which can be represented by periodic surface charge distributions of effectively infinite extent. Direct integration methods must be used for the former, but for infinite arrays the energies are calculated using Fourier methods as described in detail in Chapter 6.

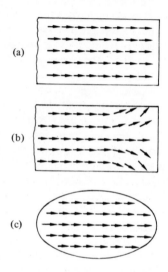

Figure 1.20. The distribution of the magnetization in a rectangular prismatic crystal with infinite anisotropy (a) and with finite anisotropy (b), and in an ellipsoid with any value of the anisotropy (c).

1.6.3 Total internal field

If an applied field, H_0, induces a level of magnetization I in an isotropic ellipsoid then, so long as $I \parallel H_0$, the total field inside the specimen will be

$$H_i = H_0 - N_d I. \qquad (1.61)$$

This simple situation can usually be achieved in practice. (If a uniformly magnetized specimen with negligible external demagnetizing factor contains an ellipsoidal hole, the total field inside the hole will be

$$H_i = H_0 + N_d I \qquad (1.62)$$

since the pole densities at the surfaces of the hole are identical in magnitude with those of a magnetized ellipsoid.)

Now suppose that an ellipsoidal specimen is divided into domains and that the domain boundaries can move without restriction (see Chapter 6). This means that the walls will move so long as the net internal field is positive, i.e. in the same direction as I, but it may be of vanishing magnitude:

$$H_0 \rightarrow N_d I$$

In this case the susceptibility, defined as the magnetization induced in the field direction per unit applied field, is given by

$$X = I/H_0 = 1/N_d. \tag{1.63}$$

An apparent flaw in this argument is that since the magnetization is no longer uniform, due to the sub-division into domains, the use of a demagnetizing factor is not justified. However, a little thought will show that, so long as the domain spacing is small compared with the dimensions of the specimen, any substantial element of the surface can be treated as a region of uniform pole density I, and the approximation is justified.

The problem of calculating magnetic susceptibilities, or curves of I against H for spontaneously magnetized specimens of real interest, is analyzed in some detail in later chapters. This becomes quite complex, but the importance of the very simple result (Equation, 1.63) for isotropic ellipsoids should be emphasized.

References

1. E. C. Stoner, *Phil. Mag.*, **23**, 833 (1937).
2. E. C. Stoner, *Phil. Mag.*, **36**, 803 (1945).
3. P. Rhodes and R. Rowlands, *Proc. Leeds Phil. Lit. Soc.*, **6**, 191 (1954).

2

Magnetic Dipoles in Applied Fields

2.1 PRECESSION

The first problem is the response of a single isolated dipole to a static applied magnetic field. It has been noted that the energy of a dipole depends upon its orientation in a field $\mathbf{H}$ according to $W = -\boldsymbol{\mu} \cdot \mathbf{H}$, and that this is a minimum $(-\mu H)$ when the dipole is aligned with the field. However, if we mean by an isolated dipole one which cannot exchange energy with its surroundings, then any motion towards the field direction is prohibited because the reduction of energy would violate the first law of thermodynamics.

To understand the way in which the dipole is, in fact, affected it is necessary to recall the origin of the dipoles of practical importance. Unlike electrostatic dipoles, magnetostatic dipoles are always associated with charges in motion; principally, as described later, with electrons orbiting around an atomic nucleus or spinning about their own axis. Thus, in turn, a certain angular momentum $\mathbf{g}$ must be associated with the magnetic dipole moment and, presuming the linear relationship, we may write

$$\boldsymbol{\mu} = \gamma \mathbf{g} \,. \tag{2.1}$$

The implication that the vectors are parallel is justified, because the angular momentum of an orbiting particle is represented by a vector directed normal to the plane in which the motion occurs. It will be assumed for the present that the magnitude of the angular momentum is unaffected, although later a correction must be made in this respect.

The torque exerted on the dipole by a field has been given as $\boldsymbol{\mu} \times \mathbf{H}$, and by Newton's laws this must be equal to the rate of change of the angular momentum. (The corresponding law for linear motion, namely that force = mass × acceleration or equals the rate of change of linear momentum, may be more familiar.) Thus, with Equation (2.1), the law of motion is

$$\boldsymbol{\mu} \times \mathbf{H} = \frac{d\mathbf{g}}{dt} = \frac{1}{\gamma} \frac{d\boldsymbol{\mu}}{dt} \,. \tag{2.2}$$

Expressing $\boldsymbol{\mu}$ in its components as in Figure 2.1, and differentiating we obtain

$$\frac{d\boldsymbol{\mu}}{dt} = \frac{d\mu_x}{dt}\mathbf{i} + \frac{d\mu_y}{dt}\mathbf{j} + \frac{d\mu_z}{dt}\mathbf{k} \tag{2.3}$$

since **i**,**j**,**k** are constant in magnitude. Clearly $d\boldsymbol{\mu}/dt$ is a vector of the familiar form.

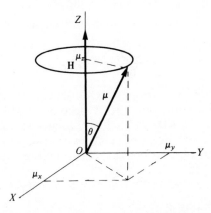

Figure 2.1. A magnetic dipole, $\boldsymbol{\mu}$, precessing about a field **H** with the coordinates such that **H** $\parallel OZ$.

The vector product, $\boldsymbol{\mu} \times \mathbf{H}$, can also be expanded into component form, using the standard determinant relation:

$$\boldsymbol{\mu} \times \mathbf{H} = \begin{vmatrix} \mathbf{i} & \mathbf{j} & \mathbf{k} \\ \mu_x & \mu_y & \mu_z \\ H_x & H_y & H_z \end{vmatrix}$$

$$= \mathbf{i}(\mu_y H_z - \mu_z H_y) + \mathbf{j}(\mu_z H_x - \mu_x H_z) + \mathbf{k}(\mu_x H_y - \mu_y H_x). \qquad (2.4)$$

If the coordinate system is chosen so that OZ is parallel to **H**, i.e. $H_z = H$, $H_x = H_y = 0$, Equation (2.4) becomes

$$\boldsymbol{\mu} \times \mathbf{H} = \mu_y H \mathbf{i} - \mu_x H \mathbf{j}. \qquad (2.5)$$

A vector equation, such as (2.2), can only be satisfied if each of the components of the two vectors are themselves equal, and thus with Equations (2.3) and (2.5) there are three scalar equations

$$\frac{d\mu_x}{dt} = \gamma \mu_y H, \qquad (2.6)$$

$$\frac{d\mu_y}{dt} = -\gamma \mu_x H, \qquad (2.7)$$

$$\frac{d\mu_z}{dt} = 0. \qquad (2.8)$$

Integrating Equation (2.8) gives

$$\mu_z = \text{constant}$$

but (Figure 2.1) $\mu_z = \mu\cos\theta$, and so this is equivalent to

$$\cos\theta = \text{constant}. \tag{2.9}$$

Differentiating Equation (2.6) and substituting for $d\mu_y/dt$ from Equation (2.7) gives

$$\frac{d^2\mu_x}{dt^2} = -\gamma^2 H^2 \mu_x.$$

This is a very familiar form of differential equation; it is solved by using the notation $D \equiv d/dt$, so that it becomes $(D^2 + \gamma^2 H^2)\mu_x = 0$, which factorises to $(D + i\gamma H)(D - i\gamma H)\mu_x = 0$ to give two first degree equations. These are solved by direct integration after separation of the variables, the most general solution being the sum of the two particular solutions, i.e.

$$\mu_x = A e^{i\gamma Ht} + B e^{-i\gamma Ht} = C\cos(\gamma Ht + \epsilon).$$

Similarly, eliminating μ_x:

$$\mu_y = C\sin(\gamma Ht + \epsilon).$$

But (Figure 2.1)

$$\mu_x^2 + \mu_y^2 = C^2 = \mu^2\sin^2\theta$$

and the final results are

$$\mu_x = \mu\sin\theta\cos(\gamma Ht + \epsilon)$$

$$\mu_y = \mu\sin\theta\sin(\gamma Ht + \epsilon) \tag{2.10}$$

$$\mu_z = \mu\cos\theta \quad (= \text{constant}).$$

The meaning of these equations becomes clear on their inspection in relation to the figure: the component parallel to **H** is always constant but the vector μ rotates to give the same components μ_x and μ_y each time γHt changes by 2π. This constitutes precession, with angular frequency

$$\omega = \gamma H.$$

It is important to note that, for this isolated dipole, there is no change in the component of the magnetization in the applied field direction. If, however, the dipole can gain or lose energy, an applied field will induce such a component.

2.2 EQUILIBRIUM DISTRIBUTION OF DIPOLE ORIENTATIONS

In contrast to the foregoing it will now be assumed that equilibrium has been attained. At a temperature of absolute zero this would simply mean that the dipole would be oriented in the field direction, but at any finite temperature T it will be subjected to thermal fluctuations of energy kT, where k is Boltzmann's constant, which tend to destroy the orientation.

The principles are the same for either a single dipole or a (non-interacting) assembly of dipoles. The component of the dipole moment, or the mean component of the moments of the assembly, per unit volume, is referred to as the induced magnetization, I, i.e.

$$I = n\langle\mu\cos\theta\rangle = n\mu\langle\cos\theta\rangle$$

where there are n dipoles in unit volume. This will clearly be a function of H and T, resulting from a balance between the aligning effects of the field and the randomizing effects of the temperature. The relevant theorem of statistical mechanics (Boltzmann's statistics) states that the mean value of any quantity, X, which may have a series of values X_j, each of which is associated with an energy w_j, is

$$\langle X\rangle \equiv \frac{\sum_j X_j \exp(-w_j/kT)}{\sum_j \exp(-w_j/kT)} ; \qquad (2.11)$$

$\sum_j$ represents summation or integration over all possible energy states. In the present case the energy term is $-\mu H\cos\theta$; other terms which are not dependent on the orientation cancel out, and thus

$$\mu\langle\cos\theta\rangle = \mu\frac{\sum_j \cos\theta_j \exp(\mu H\cos\theta_j/kT)}{\sum_j \exp(\mu H\cos\theta_j/kT)} .$$

Putting $\mu H/kT = a$, $\cos\theta_j = x_j$, and changing to the integral form (for a continuous variable)

$$\mu\langle\cos\theta\rangle = \mu\frac{\int_{-1}^{1} x\exp(ax)dx}{\int_{-1}^{1} \exp(ax)dx},$$

the limits -1 and 1 corresponding to $\theta = 0$ and 2π, for all orientations. The above integral is a standard form and gives the result

$$\mu\langle\cos\theta\rangle = \mu[\coth(\mu H/kT) - kT/\mu H] \qquad (2.12)$$

which is the Langevin function $(x\mu)$. The limit, as H becomes very large and T small, is unity. This is as it should be since it represents complete alignment or saturation, i.e. $I = n\mu$. When H/T is small the Langevin function reduces to $\mu H/3kT$, so that the induced magnetization is

$$I = \frac{n\mu^2 H}{3kT} . \qquad (2.13)$$

We now define the *susceptibility* as the ratio of the induced magnetization to the inducing field, i.e.

$$\chi = I/H.$$

The second of the above limits is then

$$\chi = \frac{n\mu^2}{3kT} = \frac{C}{T} \ .$$

$$(2.14)$$

2.3 REPRESENTATION OF CLASSES OF MAGNETIC MATERIALS

Anticipating the next chapter we state that the orbiting or spinning electrons constitute magnetic dipoles of a magnitude which is specified by quantum theory to be integral numbers of a basic unit, the Bohr magneton μ_B equal to $9 \cdot 2732 \times 10^{-21}$ erg Oe^{-1} in e.m.u. The contributions of the individual electrons sum to give values of the atomic dipole moments ranging from zero to an upper limit of the order of $10\mu_B$, as a consequence of the tendency for most of the electronic dipoles to become 'paired' and cancel out in large atoms.

When measurements of χ are made, as below, a large class of materials is found to behave in accordance with Equations (2.13) and (2.14). That is to say they have room temperature susceptibilities of the order of 10^{-5}, as forecast by inserting numerical values in Equation (2.14) (with $k = 1 \cdot 38 \times 10^{-16}$ erg deg^{-1} and $\mu \sim 1\mu_B$), and so long as measurements are made in fields of about 10^4 Oe, plots of $1/\chi$ vs. T are linear. It is only in fields approaching 10^6 Oe and at temperatures of a few degrees absolute that saturation is approached, and the saturation magnetization is given by predicted atomic moments times the number of 'magnetic atoms' or ions per unit volume. A material which shows such behaviour, as summarised in Figure 2.2, is said to be *paramagnetic* or to exhibit *paramagnetism*.

A second class, the *diamagnetic* materials, have susceptibilities which are negative, generally $< 10^{-5}$ in magnitude, and independent of temperature. The origin of the behaviour must be quite different from that considered above, and lies in an important omission made in the treatment of the effects of fields on magnetic dipoles. Continuing the classical representation of the atomic dipole as a minute current loop, and recalling Faraday's laws of induction, it is in fact to be expected that the nature of the current will itself be altered as a field is applied. A full analysis of the model shows that the radius is unchanged but, for fields applied normal to the plane of the loop, the velocity or angular frequency of the electron alters in such a way that the induced dipole field opposes the applied field. (This latter feature may be considered as a consequence of Lenz's law.) Purely diamagnetic behaviour can be represented by picturing two paired electrons as rotating in opposite directions around superimposed circular orbits: the inherent dipoles cancel out but when a field is applied the velocity of one increases and that of the other decreases, so that both contribute a negative induced dipole component. When the atoms possess dipole moments in the absence of a field, the

diamagnetic effect must be treated as a correction to the measured paramagnetic susceptibility. Since diamagnetism is independent of the pairing
of the electrons, the magnitude of the effect depends upon the total
number of electrons, z, approximately as

$$\chi_M = -10^{-8}z. \tag{2.15}$$

In this χ_M represents the induced magnetization per unit field for one
gram-atomic weight, or gram-molecular weight, of the substance. χ, as
defined earlier, referred to unit volume and χ_m is used for unit mass:
thus, with a knowledge of the density and atomic or molecular weights,
the three are equivalent.

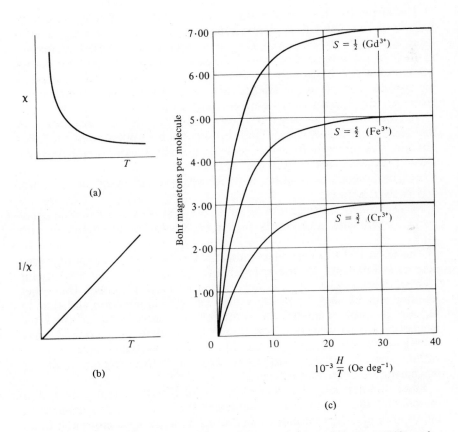

Figure 2.2. Curves (a) and (b) show the temperature dependence of the susceptibility and
its reciprocal for a simple paramagnetic material with no interactions, as measured in
moderate fields ($\sim 10^4$ Oe). Curve (c) shows real data indicating the approach to saturation
in very high fields and low temperatures (dilute salts of the ions indicated: see also
Chapters 3 and 4) [13].

Since the further development is not greatly concerned with diamagnetism, the following representative values may be noted ($-\chi$ or
$|\chi| \times 10^6$):

Zinc	0·14	Mercury	0·17	Copper	0·086
Silver	0·19	Gold	0·14	Germanium	0·106
Graphite	21·7	Graphite ($\chi\parallel$)	21·0	Graphite ($\chi\perp$)	0·3

The three values for graphite are respectively for a powder and for single
crystals measured parallel and normal to the hexagonal c-axis. The very
large value of $\chi_\parallel$ may be associated with the delocalization of the
electrons in the six-membered rings of the crystal structure, as is that of
porphyrin and other conjugated π-electron systems.

2.4 INTERNAL EFFECTIVE FIELDS

The distinctive feature of a third class of materials is that, once they
have been magnetized to saturation, they may remain saturated even
after the fields are removed (or, to a certain extent, when fields are
applied in the reverse direction). These *ferromagnetic* materials have a
behaviour which is strongly sensitive to such structural features as specimen shape, particle or crystallite size, or to the presence of inclusions.
However they always show the following features:

a) Plots of induced magnetization versus field are non-linear, even in
low fields, and irreversible. Thus, by suitable manipulation of the fields
I/H loops as shown in Figure 2.3 may be described, exhibiting hysteresis.
(Note the definitions given in the caption.)

b) The remanence may be identical with saturation, as noted above,
and is always finite.

c) The susceptibility cannot be so simply defined as for paramagnetics,
but it is generally enormously higher i.e. up to 10^5 as compared with
10^{-5}.

The feature (*b*) in particular indicates the existence of a spontaneous
magnetization, that is, the spontaneous alignment of the constituent
atomic dipoles without the necessity for applied fields. The other
features will be seen to be derivable from this supposition and thus to
support it. For example, the induction of a large component of the
magnetization in the field direction, by a small applied field, may
correspond to the rotation of the direction of the spontaneous magnetization towards the field direction, and not to the induction of the
magnetization at an atomic level.

Since dipoles may be aligned by a sufficiently strong field it is
reasonable, and very useful, to consider the spontaneous alignment of
the dipoles in a ferromagnet to be caused by a mythical molecular field,
as suggested by Weiss [1], or exchange field, H_e. (The reason for the
terminology will become apparent later.) The alignment may also be
represented by an energy relation:

$$W_e = -J\mu_1 \cdot \mu_2 \tag{2.16}$$

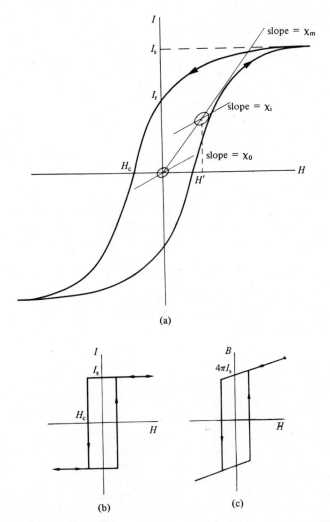

(a)

(b)   (c)

Figure 2.3. (a) A schematic I/H or hysteresis loop for a ferromagnetic (or ferrimagnetic) material, defining the following:

I_s—saturation magnetization;
I_r—remanence or remanent magnetization;
H_c—coercivity or coercive field.

The slopes of the lines indicated define

χ_m—maximum susceptibility;
χ_i—incremental susceptibility in the bias field H';
χ_0—initial susceptibility (for a minor loop of vanishingly small amplitude with no bias field, as found by extrapolation).

The differential susceptibility χ_d is the slope, at any point, of the loop itself.

(b) and (c) Illustration of the difference between I/H and B/H loops for the simple case of a square I/H loop with complete reversal at H_c. The difference is insignificant for 'soft materials' with $H_c \sim 10^{-3}$ Oe, but great for permanent magnet 'hard materials' with H_c comparable to $4\pi I_s$. For loops which are not square the fields for zero magnetization and for zero induction, i.e. $_I H_c$ and $_B H_c$, differ considerably for hard materials.

where the coefficient J is known as the exchange integral. This indicates that the interaction (or exchange) energy between two neighbouring dipoles is minimum when they are parallel, so long as J is positive. It is immediately apparent that the explanation of the alignment forces, or exchange fields, cannot be found in classical magnetostatics, since these predict an energy minimum for antiparallel dipoles.

To conform with observations it is necessary to assume that the exchange field is proportional to the magnetization itself, that is to the extent to which perfect alignment is approached:

$$H_e = \alpha I. \tag{2.17}$$

A ferromagnet can now be treated as an array of dipoles, originally disordered, on which the exchange field operates, in addition to any applied field, to produce a degree of alignment against the effect of thermal fluctuations. Thus Equation (2.12) gives, for zero applied field

$$I = n\mu L \frac{\mu H_e}{kT} = n\mu L \frac{\mu \alpha I}{kT} \tag{2.18}$$

where L represents the Langevin function. Due to the appearance of I in this function the analytical solution is difficult, but a graphical solution may be made by writing

$$I = n\mu L(x) \tag{2.19}$$

where

$$x = \frac{\mu \alpha I}{kT} \, ,$$

whence

$$I = \frac{kT}{\mu \alpha} x \, . \tag{2.20}$$

These are plotted for a range of temperatures, as in Figure 2.4a, and the intercepts at each temperature are taken to give the temperature variation of the spontaneous magnetization as in Figure 2.4b. It is clear, for example, that as $T \to 0$ the line for Equation (2.20) becomes nearly horizontal and the intersection occurs as $x \to \infty$ i.e. $L(x) \to 1$ and the dipoles are perfectly aligned:

$$I_{so} = n\mu.$$

At the other extreme, as T increases, I falls to zero when the initial slopes of the lines for the two equations are equal.

From Equation (2.20)

$$\frac{\partial I}{\partial x} = \frac{kT}{\mu \alpha}$$

and from Equation (2.19) with $x \to 0$, $L(x) \to \frac{1}{3}x$

$$\frac{\partial I}{\partial x} = \frac{n\mu}{3} \, .$$

Thus the critical temperature, the *Curie temperature*, can be related to the exchange field coefficient α as

$$T_C = \frac{n\mu^2\alpha}{3k} .$$ (2.21)

Practical values of T_C indicate that the order magnitude of the exchange field is 10^7 Oe. Thus, the effect of laboratory applied fields, on the magnitude of I_s, is very small.

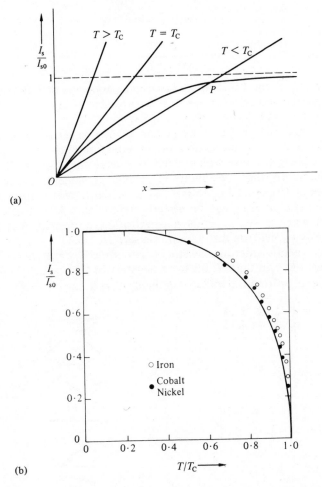

(a)

(b)

Figure 2.4. (a) Illustration of the graphical solution of Equations (2.19) and (2.20). $T > T_C$ always gives $I = 0$. For $T < T_C$ the intercepts increase as T falls indicating a regular rise in I (or I_s, the spontaneous magnetization) towards the limiting value I_{s0} representing complete alignment of the dipoles. The resulting solution for I_s versus T is shown in (b), with experimental data [14].

Above T_C the ferromagnet behaves as a paramagnetic material, and magnetization can only be induced by an applied field. However, there is no reason to suppose that H_e disappears: although it can no longer cause alignment on its own it increases the susceptibility in an applied field. Replacing H by $(H+\alpha I)$ in Equation (2.13) gives

$$I = \frac{n\mu^2(H+\alpha I)}{3kT} .$$

Solving for I we get

$$I = \frac{n\mu^2 H}{3kT - n\mu^2\alpha} = \frac{n\mu^2 H}{3k(T-T_C)}$$

giving

$$\chi = \frac{n\mu^2}{3k(T-T_C)} . \tag{2.22}$$

The observation of the Curie–Weiss law for a paramagnetic indicates the likelihood of its becoming ordered at a sufficiently low temperature, although this does not always follow.

For some materials a Curie–Weiss law with a negative value of T_C is observed. This is an indication of an interaction of the opposite sign to the above, i.e. one which tends to reduce the susceptibility by torques tending to align neighbouring atomic movements in an anti-parallel manner (Figure 2.5). Indeed, at sufficiently low temperatures, antiferromagnetic ordering can be directly demonstrated: not of course so easily as when a spontaneous magnetization is present, but quite unambiguously by neutron diffraction (see section 2.6.4).

Antiferromagnetic ordering may still be represented by Equation (2.16), but the exchange constant, J, must be negative in this case, to indicate a minimum energy for antiparallel alignment ($\theta = \pi$, $\cos\theta = -1$).

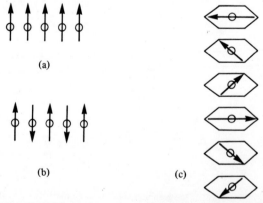

Figure 2.5. The arrangement of the ionic magnetic moments (spin components) in a ferro-magnetic (a) and in antiferromagnetics (b and c) with zero spontaneous magnetization. (Perfect order is assumed, as at very low temperatures.) The spiral structure (c) is found in some rare-earth metals, such as holmium between 20°K and 132°K.

The interactions are most clearly demonstrated by plotting $1/\chi$ versus T as in Figure 2.6. Inverting Equation (2.22) gives

$$\frac{1}{\chi} = AT - B \qquad (2.23)$$

where A and B are constants. A positive exchange interaction corresponds to a positive value of B and vice versa, giving the equally-spaced lines shown in the figure for the same value of T_C. For antiferromagnetics the interaction term is often represented by θ rather than T_C. T_C for ferromagnetics is the Curie point, below which the ordering becomes effective according to the simple treatment (although this is, in fact, an approximation and there is usually a slight difference between the Curie point and Weiss constant) whereas since θ is a negative temperature, it cannot also represent the temperature below which the antiferromagnetic ordering is effective. This latter is known as the Néel temperature, T_N. The existence of a maximum in χ, as plotted against T, may be taken to demonstrate the existence of antiferromagnetism and as an indication of the Néel temperature.

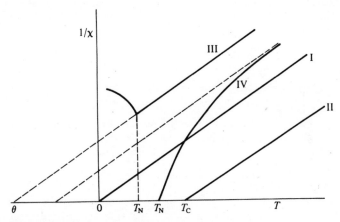

Figure 2.6. The reciprocal of susceptibility versus temperature:
   I—Simple paramagnetic with no interactions: Curie Law.
   II—Curie-Weiss paramagnetic with positive interaction term T_C, indicating ferromagnetic ordering below T_C (the Curie temperature).
   III—Antiferromagnetic with negative interaction constant, θ. T_N is the Néel temperature below which the interaction is effective in causing antiferromagnetic ordering. χ falls below T_N because the moments are effectively locked into a particular orientation in the lattice and an applied field must overcome this anisotropy as well as the ordering.
   IV—Ferrimagnetic.

The line marked IV can be shown to correspond to the particular type of ordering, which is very similar to antiferromagnetism but yet gives a net spontaneous magnetization, known as ferrimagnetism. This arises when two different sets of atomic or ionic magnetic moments form interpenetrating lattices: each is ordered in a ferromagnetic manner but the mutual orientation of the two sub-lattices is of the antiferromagnetic type, as shown in Figure 2.7. A spontaneous magnetization exists, below the critical temperature, if the two types of magnetic moments differ in magnitude, as shown in Figure 2.7a, or in number.

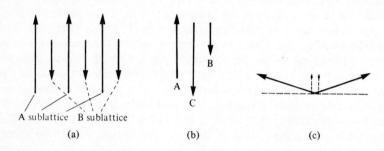

Figure 2.7. Moment arrangements in ferrimagnetics: (a) simple type occurring in spinel ferrites; (b) three sub-lattices as in rare-earth iron garnets; (c) canted spin arrangement as in orthoferrites or haematite, giving a weak resultant moment normal to the ordering direction.

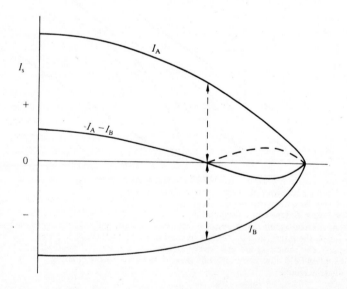

Figure 2.8. If the magnetization of the sub-lattice A with the higher value of I_{s0} falls the more rapidly with rising temperature, the resultant spontaneous magnetization may pass through zero at a compensation temperature before finally falling to zero at the Néel temperature. When measurements are made in an applied field the results indicated by the broken line are obtained.

As a first approximation the two sub-lattices of a ferrimagnetic material may be treated independently and, if each has a different rate of change of magnetization with temperature, there is the possibility of finding two temperatures at which the spontaneous magnetization falls to zero. This is explained by Figure 2.8; the first temperature at which $I_s = 0$ is known as the compensation temperature. Figure 2.9 gives data for garnets which show this effect.

It should be stressed that the whole of the above treatment has been made for an arbitrary model of a collection of dipoles, with interactions represented by effective fields and with no very direct reference to real materials. It is quite remarkable that this treatment does correspond to the real behaviour of materials to a considerable extent, but not surprising that there are exceptions. The application to simple ionic solids is very good but the behaviour of metals may diverge considerably from that predicted, corresponding to the applicability of Fermi–Dirac rather than Boltzmann statistics [2,3]. Thus the presence of a positive exchange interaction at elevated temperatures does not *always* correspond to ferromagnetic ordering at lower temperatures, and over a range of some hundreds of degrees centigrade the susceptibility may be almost independent of temperature (see section 4.3).

2.5 ANISOTROPY FIELDS

According to the representation of a ferromagnetic as given so far, it follows that if a specimen is rotated in an applied field the magnetization should always remain precisely in the field direction. In practice there is always a tendency for the dipoles to remain oriented along one or more specific directions in the crystal lattice, the 'easy directions'. In cobalt there is a single easy axis (two directions) which is the principal axis of the hexagonal crystal structure; for iron the easy axes are $\langle 100 \rangle$ and for nickel and many ferrites $\langle 111 \rangle$. The tendency for the magnetization to lie parallel to the easy axis may be represented by an anisotropy energy which is zero for perfect alignment and, as a result of experiment, varies with the angle, θ, between the axis and I_s as (per unit volume)

$$W_a = K_1 \sin^2\theta + K_2 \sin^4\theta \qquad (2.24)$$

plus higher even-order terms in θ. This applies to uniaxial materials while for cubic materials the direction cosines $\alpha_1, \alpha_2, \alpha_3$, i.e. the cosines of the angles which I_s makes with the cube edges, must be invoked and the corresponding law is

$$W_a = K_1(\alpha_1^2\alpha_2^2 + \alpha_2^2\alpha_3^2 + \alpha_3^2\alpha_1^2) + K_2\alpha_1^2\alpha_2^2\alpha_3^2 \qquad (2.25)$$

plus higher order terms, which are not usually taken into account.

Because of the way in which the direction cosines are defined, the first term of W_a is minimum with $I_s \parallel \langle 100 \rangle$ if K_1 is positive, and minimum with $I_s \parallel \langle 111 \rangle$ if K_1 is negative. Thus a positive K_1 applies to crystals

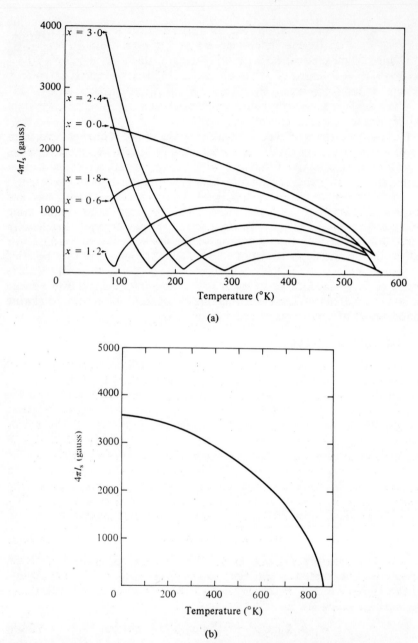

(a)

(b)

Figure 2.9. (a) The spontaneous magnetization of gadolinium-substituted yttrium iron garnet ($Gd_xY_{3-x}Fe_5O_{12}$), as a function of temperature, showing the compensation effect. Note that pure gadolinium iron garnet ($x = 3$) becomes non-magnetic near to room temperature. Similar effects are found for a few spinel ferrites, but not the common single ferrites of Mg, Mn, or Ni: results for polycrystalline $NiFe_2O_4$ are given in 9(b) [15, 16].

such as iron, and a negative K_1 to nickel. The magnetocrystalline anisotropy may also be represented as a torque on the magnetization, which is given by the rate of change of energy with respect to the angle involved. For small values of θ only the first term of Equation (2.24) need be considered and thus

$$T = \frac{\partial W_a}{\partial \theta} \doteq 2K_1 \sin\theta \ .$$

We now introduce an effective anisotropy field $\mathbf{H_K}$, which is defined as that field which would have the same effect as the crystal anisotropy i.e. give the same torque so that $\mathbf{H_K} \times \mathbf{I_s} = H_K I_s \sin\theta = 2K_1 \sin\theta$:

$$H_K = 2K_1/I_s \tag{2.26}$$

with the obvious implication that $\mathbf{H_K}$ is directed along an easy direction. It can be shown that the same expression applies to cubic crystals with positive K_1 (I_s nearly parallel to $\langle 100 \rangle$) while for negative K_1

$$H_K = -\frac{4K_1}{3I_s} \quad \left(= \frac{4|K_1|}{3I_s} \right) . \tag{2.27}$$

The ferromagnetic elements, iron cobalt and nickel, have the following values of K_1 and H_K:

	K_1 erg cm^{-3}	H_K Oersted
iron	$4 \cdot 8 \times 10^5$	560
cobalt	53×10^5	7400
nickel	$-0 \cdot 45 \times 10^5$	185

Materials with much larger values of H_K are discussed in Chapter 5.

It is also noted that, apart from crystal anisotropy, two other effects may influence the direction of the spontaneous magnetization in the absence of an applied field. The basis of the first of these has already been given, that is the dependence of the demagnetizing energy of a non-spherical particle on the direction of the magnetization. This gives rise to a shape anisotropy which tends to align the magnetization with the major axis of a wire for example. The second of these additional effects arises from *magnetostriction*.

It is observed that the shape of most ferromagnetic specimens changes during the course of magnetization and this magnetostrictive effect may be measured by a single coefficient for a polycrystalline material at saturation

$$\lambda_s = \frac{\delta l}{l} \qquad (\text{at } I = I_s) .$$

Conversely, if a stress, σ, is applied to a ferromagnetic specimen, there is found to be an interaction between the stress direction and the direction $\mathbf{I_s}$ which is indicated by an energy term dependent on the mutual orientation of the two, θ, as (per unit volume)

$$W_\sigma = -\tfrac{3}{2}\lambda_s \sigma \sin^2\theta, \tag{2.28}$$

and the effect may also be regarded as a stress anisotropy with an effective coefficient $\frac{3}{2}\lambda_s\sigma$. Thus the stress may be taken to modify the crystal anisotropy, replacing K_1 by $(K_1+K_\sigma) = K_1+\frac{3}{2}\lambda_s\sigma$. This is an approximation, since magnetostriction is highly anisotropic with respect to the crystalline directions to which it is applied: e.g. for iron

$$\lambda_{100} = 20\cdot7 \times 10^{-6}, \qquad \lambda_{111} = -21\cdot2 \times 10^{-6},$$

and for nickel

$$\lambda_{100} = -45\cdot9 \times 10^{-6}, \quad \lambda_{111} = -24\cdot3 \times 10^{-6}.$$

For single crystals such values should be used together with the elastic moduli for the materials.

For polycrystalline specimens with random orientation

$$\lambda_s = \tfrac{2}{5}\lambda_{100}+\tfrac{3}{5}\lambda_{111},$$

and values of λ_s are:

iron, 7×10^{-6};   nickel, 34×10^{-6};   cobalt, 50×10^{-6}.

Inserting reasonable values for σ (below the breaking stress or yield point) it may be seen that for nickel the stress anisotropy can easily exceed the crystal anisotropy: such materials may be considered to be very strain-sensitive.

2.6 MAGNETIC MEASUREMENTS

The very great number of different methods of measuring magnetization may be divided crudely into four categories.
1. Measurement of dipole fields.
2. Measurement of induction using coils.
3. Measurement of the force on a specimen in a non-uniform field.
4. Less direct measurements.

Since a comprehensive review is impossible, a rather personal view may be permitted, and a brief account will be given of methods used or developed by the author.

2.6.1 Dipole field measurements

Figure 2.10 shows a magnetic specimen placed centrally in a pair of Helmholtz coils, or between the pole faces of an electromagnet. It is found that, if the coils have a radius which is equal to their spacing, the field produced is uniform over a fair volume of space. Thus the field configuration can be represented by the superimposition of a uniform field and a dipole field from the specimen. If the specimen is spherical, it may be represented by a point dipole of strength vI at its centre; non-spherical specimens will generate fields of the point dipole form at points distant from the specimen compared with its dimensions.

Since the dipole fields at any point (r,θ) or (x,z) in a plane containing the dipole at the origin can be specified by the components:

$$H_r = \frac{2M\cos\theta}{r^3}$$

$$H_\theta = \frac{M\sin\theta}{r^3}$$ (2.29)

where $M = vI$, it is only necessary to measure this field to know the relative value of the induced magnetization I. However the measurement should be independent of the applied field. This can be achieved by using a Hall probe, that is, a semiconducting crystal which gives a measurement of the component of the magnetic field across its surface due to the deflection of the paths of the conduction electrons. If the connections are made appropriately, and the crystal is placed as shown in the figure, it will detect only the component H_z of the dipole field.

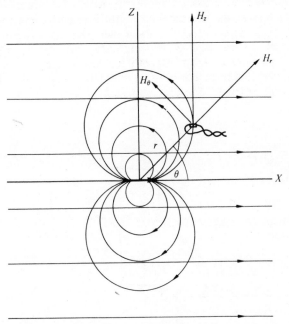

Figure 2.10. The dipole fields from a magnetized specimen ($\rightarrow$) are shown superimposed on a uniform applied field, with a Hall probe positioned so as to measure the dipole field while being insensitive to the applied field.

For any displacement of the probe along the field direction, x, the optimum displacement, z, is easily found. From Equation (2.29)

$$H_z = \frac{2M\cos\theta\sin\theta}{r^3} + \frac{M\sin\theta\cos\theta}{r^3} = \frac{3M\cos\theta\sin\theta}{r^3}$$

Considering x as constant, put $r = x/\cos\theta$ whence

$$H_z = \frac{3M\cos^4\theta \sin\theta}{x^3} \; .$$

The maximum value of H_z occurs when $\partial H_z/\partial\theta = 0$, i.e.

$$\frac{3M(\cos^5\theta - 4\cos^3\theta \sin^2\theta)}{x^3} = 0 \; ,$$

$$\tan\theta = \frac{z}{x} = \frac{1}{2} \; .$$

Thus the optimum position of the probe is at a height equal to half the distance from the dipole measured along the field direction. It is only necessary to feed the signals from the Hall probe and the current through the coils to an X/Y recorder, to plot loops such as that shown in Figure 2.11 semi-automatically. The example is for a small, weakly magnetic crystal of europium orthoferrite with a total saturation moment of only $0 \cdot 1$ e.m.u., so the method is quite sensitive even when using a commercial Hall gaussmeter [4]. However, the main reason for this full description of the method is its simplicity and the insight it can give into field distributions from magnetized specimens. With appropriate procedures, which the reader may be able to conceive, absolute measurements of I_s can be made, and H_K estimated.

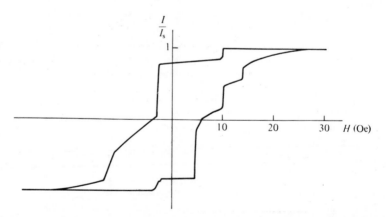

Figure 2.11. Example of a magnetization loop, measured by the arrangement of Figure 2.10, for a crystal of europium orthoferrite of volume $7 \cdot 5 \times 10^{-2}$ cm^3, $I_s = 6 \cdot 7$ e.m.u. cm^{-2}.

In a more established method, with sensitivity adequate for paramagnetic and diamagnetic susceptibilities, the specimen is vibrated along the OZ axis and the spatially fluctuating dipole fields induce small voltages in coils with their planes parallel to H_0 (i.e. OX). Stationary search coils cannot, of course, detect static dipole fields. A small permanent magnet, or coil, mounted rigidly with respect to the specimen, can be used to generate a reference signal, so that a latch-in amplifier or phase-sensitive detector can be used and the measurement be made to be independent of drifts in either the frequency or amplitude of the vibrations. Commercial equipment is expensive and even with skilled help a really high sensitivity may only be achieved by prolonged individual development work. Reference to the original papers [5,6] shows, however, that the inherent advantages of the method justify such effort. Figure 2.12 shows how the method can be adapted to study interfacial effects by maintaining the specimen in readily controlled atmospheres, without having to transmit the vibrations through a gas seal.

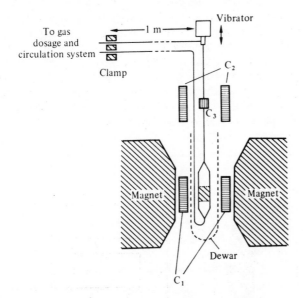

Figure 2.12. A vibrating magnetometer arranged for the study of the effects of adsorption, oxidation etc., on the saturation magnetization of powders: for this the glass tube enclosing the specimen is caused to vibrate as a whole. C_1—pickup coils, C_2—compensation coils connected in opposition to C_1, C_3—a small coil carrying a steady current which generates a dipole moment equal to that of the specimen and vibrates with equal amplitude and frequency.

2.6.2 Induction methods

A strip of high-purity iron, for example, may have such a high inherent susceptibility that, if it were isolated, the properties would be largely controlled by demagnetizing effects. However, these effects can be substantially suppressed by placing the strip in a yoke of material with similar susceptibility but larger cross-sectional area, as indicated in Figure 2.13. Since there are now no stray fields, the above methods cannot be used.

If a coil is wrapped around the specimen, and the induction is changed by applying a d.c. field (using a second coil lying within the first), a voltage v will be generated in the search coil which is proportional to the rate of change of the induction. To obtain the induction change itself it is necessary to integrate, i.e.

$$\Delta B = \int_0^\infty v \, \mathrm{d}t \ ,$$

and this is the principal problem, since the specimen should be thin to avoid residual demagnetizing effects and even for $\Delta B \to 4\pi I_s$ the voltages per turn of pick-up coil are small. In the author's experience the best low-drift integrator consists of a voltage-to-frequency converter, typically drifting by no more than 1 pulse per second and giving 10 p.p.s. for $10 \, \mu V$, and a gated counter followed by a printer or tape punch. With a large input impedance a great number of turns may be used, and it is quite practicable to measure changes in the magnetization of a strip or wire of iron of $0 \cdot 1 \, cm^2$ section, which correspond to less than one part in 10^4 of saturation. If the change of magnetization can be made in less than one second, the sensitivity is even higher than this.

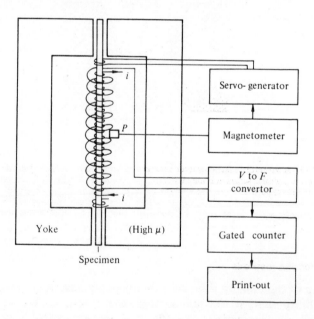

Figure 2.13. The probe P detects residual demagnetizing fields which are not quite eliminated by the mu-metal yoke: these are compensated by the small coils at the ends of the specimen, activated by the servo-loop. The narrow coil provides the applied fields and the larger coil detects the changes in induction.

2.6.3 Force methods

The energy of a specimen with permanent magnetic moment $v\mathbf{I}_s$ in a field $\mathbf{H}$ is $v\mathbf{I} \cdot \mathbf{H}$. If the field is non-uniform, the specimen is subject to a force

$$F = -\text{grad}\,(v\mathbf{I}_s \cdot \mathbf{H})$$

or

$$F = v I_s \frac{\partial H}{\partial x},$$

if the specimen is assumed to be aligned by the field and so constrained that only the component of the force in one direction need be taken into consideration. For a paramagnetic or other material in which the magnetization is induced by the field itself, the factor $\frac{1}{2}$ must be introduced and

$$F = \tfrac{1}{2} v \chi \frac{\partial (H^2)}{\partial x} = v \chi H \frac{\partial H}{\partial x}.$$

Thus permanent, or induced, magnetization (i.e. susceptibility) may be measured by the arrangement shown in Figure 2.14. The magnetic force is recorded as an effective change of weight, which is negative for a diamagnetic material, and the sensitivity depends on that of the balance. By using suitable pole pieces it is possible to obtain values of $H\dfrac{\partial H}{\partial x}$ of about 10^6, which are constant over about 1 cm. A good electro-microbalance resolves 1 μg, and in addition has the necessary characteristic of maintaining the specimen in the same place. Putting $F = 10^{-6}\,\text{g} \doteqdot 10^{-3}\,\text{dyne}$, $v = 1\,\text{cm}^3$, the limiting resolution corresponds to $\chi = 10^{-9}$ and so there is no lack of sensitivity, even if v is set at a much smaller value. The absolute accuracy, however, depends on a number of factors such as the field and field-gradient calibrations, and the reproducibility of specimen position.

This latter difficulty can be overcome by placing the specimen inside a coil, through which a current is passed to generate a magnetic moment equal and opposite to that induced in the specimen (Figure 2.14b). In this condition the net force is zero and the induced magnetization is read in terms of the compensating current, giving a result which is independent of the value of the field gradient [7]. A simple magnetometer constructed by the author gives adequate readings for about 100 mg of paramagnetic or diamagnetic specimens in fields as low as a few hundred oersteds.

2.6.4 Indirect methods of study

Only very brief reference to other methods of study can be given.

Anisotropy can be measured in terms of the torque exerted on a specimen when the applied field is rotated (as well as from magnetization curves). The coefficients are determined by fitting a theoretical curve to the measured torque curve for 180° rotation. For specimens with very high crystalline or shape anisotropy (thin films) torque measurements (that is measurement of **M** × **H**) can be used to determine the magnetic moment and saturation magnetization.

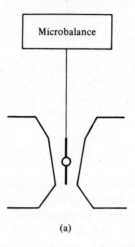

(a)

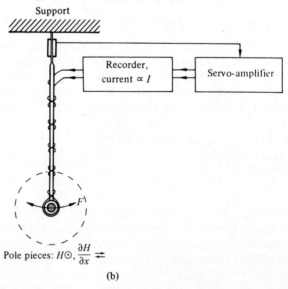

(b)

Figure 2.14. (a) Principle of the force method for magnetization or susceptibility: extra force ↓ for paramagnetics, ↑ for diamagnetics. (b) Pendulum method: the displacement due to the field gradient (field normal to diagram, field gradient left to right) may be detected by a strain gauge attached to the flexible strip and balanced out by a current passed through the coil around the specimen.

Both anisotropy and magnetization can be determined by ferro-magnetic resonance (see section 5.11).

When a specimen with no magnetic order contains nuclei which have magnetic moments, the application of a static field causes those moments to precess at a frequency proportional to the field. This can be detected, because it is the same frequency at which energy can be absorbed from an oscillating field. In a ferromagnetic specimen resonant absorption can be obtained at a frequency related to a 'contact field' due to the ordered electron spins. This provides a method, for example, of studying Curie points without the necessity of applying fields as in a conventional magnetometer. This is important, since applied fields themselves affect Curie points to a limited extent.

The magnitude and arrangement of individual atomic moments can sometimes be inferred from measurements of magnetization and anisot-ropy (including antiferromagnetic anisotropy etc.), but in this respect neutron diffraction is far more powerful. Neutrons with appropriate energy are diffracted by crystals in much the same way as X-rays, but

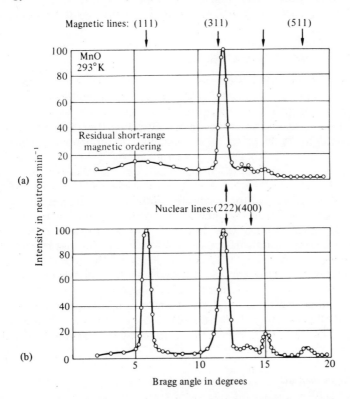

Figure 2.15. Neutron diffraction spectra for antiferromagnetic MnO, above (a) and below (b) the Néel temperature [17].

neutrons have magnetic moments and thus 'see' a different lattice according to whether it contains ordered or disordered atomic moments of different kinds. For example, Figure 2.15 shows neutron diffraction spectra of MnO above and below the Néel point. The dimensions of the magnetic unit cell are double those of the chemical unit cell, and from such results it can be shown that the magnetic moments of the Mn^{2+} ions on any one (111) plane are all parallel but that they reverse from one plane to the next. An enormous amount of detailed information on complicated ferrimagnetic and antiferromagnetic arrangements has been obtained in this way. Remarkably detailed information may be obtained on the atomic moments in metals, including the separate contributions of electron spin and orbital motion when polarized neutron beams are employed [8]: in section 4.3.3 some results for nickel are given, the spin distribution being shown by Figure 4.14.

Electron spin resonance (see section 3.12) is one of the most powerful methods for the investigation of paramagnetics at the atomic level. Fundamental information on the quantum states of the ions can be obtained and, although the analysis may be difficult, several of the uncertainties associated with the interpretation of susceptibility data are obviated.

2.7 TECHNICAL MAGNETIC MATERIALS

One object of this book is to provide a background for the study of magnetic materials. It is convenient to survey the subject briefly at this stage, though necessarily with many references ahead.

All materials have characteristic magnetic properties, but generally a 'magnetic material' is taken to be one having a spontaneous magnetization, that is, either ferromagnetic or ferrimagnetic. The representative ferromagnetics are the elements iron, cobalt, and nickel (section 4.2.2) and alloys of these with each other (such as nickel–iron or Permalloy) and with other elements (for example platinum–cobalt, silicon–iron). Some alloys of 'non-magnetic' metals are also ferromagnetic; few of these have technical applications, but with some important exceptions such as manganese bismuthide (see section 5.7). The representative ferrimagnetics are the ferrites, spinel oxides with the general formula $MeFe_2O_4$ or $MeOFe_2O_3$ where Me is a divalent ion (manganese, nickel, magnesium etc.) and Fe represents trivalent (ferric) iron; these have also been mentioned in section 2.3 and are described further in section 4.2.1.

The applications of magnetic materials are so diverse that classification is difficult. The division between those used as permanent magnets and the remainder is reasonably clear.

2.7.1 Permanent magnets

A permanent magnet consists ideally of a material in which the spontaneous magnetization is uniform, corresponding to saturation of the specimen as a whole, and is unaffected by any fields to which the magnet might be exposed in use. The function of a permanent magnet is to produce magnetic fields, without generating or driving currents, and clearly a high value of the spontaneous magnetization is desirable. The requirements are discussed more specifically in section 5.6.

The simplest representation of a magnetic material is one in which the spontaneous magnetization is maintained in a fixed orientation, with respect to a crystallographic easy direction, by anisotropy fields. From this point of view all magnetic materials could constitute useful permanent magnets so long as the anisotropy fields are high. Ideally, after saturation in a field applied parallel to an easy direction, the magnetization should be completely stable with $I_r = I_s$ in zero applied field (Figure 2.3b), provided that the specimen is spherical or ellipsoidal in shape so that the demagnetizing fields are uniform. Assuming that the only contribution to the anisotropy is of magnetocrystalline origin (that is neglecting effects of shape, or stresses) the magnetization should remain uniform in direction and magnitude even in anti-parallel fields, until these reach the value $2K/I_s$. Thus the coercive, or switching, field in which the magnetization rotates through $180°$ (section 5.1) is also equal to the anisotropy field. As noted above it can have very high values.

As explained in section 5.9 this simple behaviour in fact only arises in restricted conditions, specifically in very small crystals. It should apply to large crystals also, but in practical (imperfect) crystals domains are formed (nucleate) in fields which are well below $2K/I_s$ and may be less than the demagnetizing field of the saturated specimen. However, the principle remains that any magnetic material can have a high coercivity associated with crystal anisotropy, or with shape anisotropy if K/I_s is not high, when in a suitable physical form. To this extent there are no unique 'permanent magnet materials', only certain materials which can conveniently be prepared with an appropriate structure, that is one in which domains cannot nucleate and the anisotropies have their full effects.

2.7.2 High permeabilities (low frequencies)

Once domain walls are introduced low coercivities and high permeabilities can readily be explained. At this stage it is only necessary to note that a domain wall represents a transitional layer between two regions or domains in which the spontaneous magnetization differs in orientation, and that it has a finite energy and width, generally much less than the domain width. Details are given in section 6.3.

In a perfect crystal the energy of a domain wall is independent of its position. Thus the wall in a two-domain crystal, for example, is expected to move spontaneously to a position which minimizes the

demagnetizing energy (Figure 2.16). This is the position which bisects
the particle, which is thus demagnetized and has zero coercivity and
zero hysteresis loss. In the presence of the wall, assuming no barriers
to nucleation or that saturation is never quite achieved in a magnetization cycle, finite coercivity and hysteresis arise only from structural
imperfections (section 6.4).

In an applied field the domain magnetized parallel to the field will
grow at the expense of the other, since this decreases the magnetic
energy term $(-\mathbf{I} \cdot \mathbf{H})$. The growth will be limited by the increase in
demagnetizing energy, which is least when the wall is central and greatest
when it approaches one edge. The susceptibility approximates to $1/N$, at
least in ellipsoidal particles for which the demagnetizing factor can be
simply defined. Thus, very high susceptibilities can arise by combining a
low demagnetizing factor with a high degree of crystal uniformity. Low
anisotropies are also desirable since, even when bulk rotation of the
magnetization is not involved, this reduces the effect of imperfections.

Good crystal uniformity corresponds to high purity, freedom from
inclusions, crystal faults, stresses etc. Low effective demagnetizing factors are obtained by producing specimens in the form of rings, thin
sheets which fit together to form continuous frames, or thin tapes
or ribbons which can be rolled into toroids (Figure 2.17). In practice
it is possible to obtain coercivities well below $0 \cdot 1$ Oe, and values of
$\mu_m > 100\,000$; exceptionally, greater than $1\,000\,000$. (It is recalled that
$\mu = 4\pi\chi$ effectively, for high permeability materials).

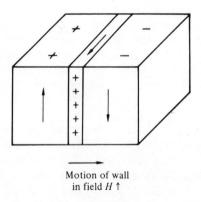

Motion of wall
in field $H \uparrow$

Figure 2.16. A single domain wall dividing a crystal into two domains of uniform
magnetization. The net magnetization is zero, but a resultant magnetization arises when a
field is applied and the domain wall is displaced as indicated.

2.7.3 The utilization of high-permeability materials

The self-inductance L of a coil is the constant of proportionality between the flux produced by the coil, that is $N = \int \mathbf{B} \cdot d\mathbf{S}$, and the current i flowing through it:

$$N = Li.$$

If the coil is wound in the form of a ring-shaped solenoid enclosing a ring of magnetic material of permeability μ, then its inductance will clearly be proportional to μ, since $\mu = B/H$ where H is the field produced in air. Consequently, self-inductances are enormously increased by enclosing high-permeability materials in the coils.

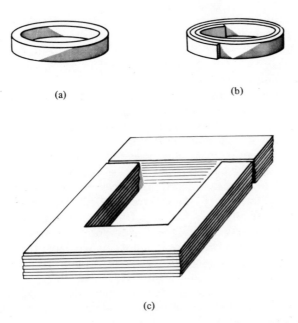

(a)   (b)

(c)

Figure 2.17. Practical specimen shapes with low demagnetizing factors, typically (a) solid ferrite ring core, (b) nickel–iron (Permalloy) tape wound core and (c) transformer core of laminated silicon–iron sheets.

Mutual inductance M between two coils can readily be shown to be increased by the same factor, since it is given by

$$N_2 = M_{21}i_1,$$

where N_2 is the flux through a coil 2 due to a neighbouring coil 1 carrying a current i_1. If both coils are wound on the same ring of material the flux enclosed by 2 is given by $\int \mu \mathbf{H} \cdot d\mathbf{S}$, where $\mathbf{H}$ is the field produced in the absence of the magnetic material.

Faraday's laws of electro-magnetic induction state that

1. when the flux through a circuit changes an e.m.f. is induced in the circuit;

2. the e.m.f. is proportional to the rate of change of flux, that is

$$V = \int \mathbf{E} \cdot d\mathbf{l} = -\frac{d}{dt} \int \mathbf{B} \cdot d\mathbf{S}.$$

Thus, when a current i_1 is applied to the coil 1, the voltage induced in 2 is proportional to the permeability of the material enclosed by the coils (for a constant rate of change, or frequency, of the current). The requirements for voltage transformation are thus, again, a high permeability combined with a suitable design such that all the flux generated by one coil passes through the other.

The technical importance of transformers in the transmission of electrical energy is enormous. Recalling the rôle of magnetic materials in dynamos and motors for the generation and utilization of electrical energy, it is easy to comprehend that more than one million tons of such material is produced annually.

For high power transformers the core material is driven close to saturation at each cycle, that is a major hysteresis loop is described 50 times per second. Thus it is the maximum permeability, μ_m, which is of importance and in practice this must be combined with economic considerations. Furthermore, the losses to be taken into account are the hysteresis loss and the eddy-current loss: silicon iron (see sections 6.5.3, 6.6.1) is found to have reasonably low hysteresis and, when in the form of laminations about $0.01''$ thick, low eddy-current losses, and now is the material universally used for large transformers. High quality transformers, however, may utilize the higher permeabilities of nickel–iron alloys (section 6.5.2). These are much more malleable than silicon iron, particularly when other elements such as copper are included in the composition, and can be rolled into very thin strips. The strips are wound into toroidal cores which have minimal eddy current losses (compared with the frames built up of relatively thick and rigid laminations of silicon iron; see Figure 2.17).

The losses in transformers, which appear as heat, still constitute a major problem equivalent to the waste of many millions of pounds worth of power per year. This gives a great incentive to the further understanding of the principles involved, and explains why very considerable research efforts are still devoted to such long-standing problems as hysteresis and eddy-current losses.

2.7.4 High-frequency applications

In high frequency applications, as in the receivers used in telecommunication, the most representative problem is that of amplifying a signal of very small amplitude. Thus, in contrast to low-frequency power engineering, the most important parameter is the initial permeability, μ_0. This may be as high as 100 000 in complex alloys based on nickel–iron.

However, the specification of a single value of μ_0, by implication that measured at low frequencies, is insufficient since it is always found that the permeability is a function of frequency. Furthermore, at high frequencies, the induced magnetization does not remain in phase with the driving field and thus the permeability should be taken as being complex

$$\mu = \mu' + i\mu'',$$

so that if the driving field is given by $H = H_0 e^{i\omega t}$ and the induction represented by $B = B_0 e^{(i\omega t + \delta)}$, then the losses can be represented by the lag or loss angle δ, where

$$\tan\delta = \frac{\mu''}{\mu'}.$$

An inverse indication of the losses is the quality factor or magnification factor of an inductor

$$Q = \frac{1}{\tan\delta} = \frac{\mu'}{\mu''}.$$

The range of useful frequencies for a particular material in a specified form can thus only be indicated by plots of the real and imaginary parts of the permeability, μ' and μ'', against frequency (permeability spectra) or by plots of Q against frequency (see sections 5.10, 5.11 and 6.7).

An even more general indication of the usefulness of an inductor core material can be given in terms of the ratio of the loss angle to the real part of the initial permeability (that part which gives the in-phase magnetization), that is the quantity $\tan\delta/\mu'$. The most useful material at any given frequency is that with the lowest value of this ratio. Table 2.1 indicates that ferrites 'overtake' even thin alloy tapes at quite moderate frequencies, from this point of view.

Table 2.1.

Frequency (kHz)	$(\tan\delta)/\mu' \times 10^6$	
	Alloy tapes	Ferrites
10	10	5
100	50	10
1 000	5 000	100
10 000	–	200

The values are only approximate: those for the ferrites represent the most suitable materials at the particular frequencies; manganese–zinc ferrite up to 1 MHz and nickel–zinc ferrite above this.

One way of explaining the different ranges of application is to note that, at kilohertz frequencies and below, μ'' will generally be low and $\mu_0 \doteq \mu'$ is the primary parameter, so long as the eddy currents are inhibited by a low tape thickness. Alloys are then superior to manganese zinc ferrites. At increasing frequencies eddy-current losses become increasingly important, causing μ'' to rise rapidly for the alloys, there being a limit to the minimum thickness to which the tape can be reduced. Manganese zinc ferrite then becomes more useful, at about 10 kHz because, being an oxide, it is a relatively poor conductor. (Ferrites are used in the solid form, not as laminations or powders.)

Unfortunately manganese–zinc ferrite owes its relatively high (low frequency) permeability to a very small crystal anisotropy [9], which in turn can only be achieved with a large concentration of ferrous ions and this renders it a poor insulator in comparison with some other ferrites. Thus the eddy-current losses, although small, eventually become prohibitive at increasing frequencies. Then nickel–zinc or other ferrites, with resistivities of the order of 10^7 ohm cm, must be used, although they have low frequency values of μ_0 or μ' which are much lower than those for other materials.

It will be seen that even when eddy-current losses are completely negligible (and hysteresis losses are quite generally neglected where only the initial permeability is concerned) the range of useful frequencies will still be limited by the losses associated with resonance (section 5.11). Also, certain basic conflicts are found to arise between a high value of μ_0 and a high resonance frequency, as explained in chapter 5.

2.7.5 Microwave applications

In a representative microwave application a sheet or rod of ferrimagnetic material, located in a cavity or waveguide, is usually magnetized to saturation and interacts with electromagnetic radiation of cm or mm wavelength. Advantage is then taken of the phenomenon of resonance: if the material has no medium frequency loss mechanisms, that is it is a good insulator so that there are no appreciable eddy currents, and is also free from ferrous ions which give rise to losses by an electron exchange mechanism [10], then the absorption of energy should occur only at and around a resonant frequency ω_0. This implies that the losses are inappreciable at frequencies distant from ω_0, and μ'' rises to a peak at ω_0.

Neglecting the effects of the crystal anisotropy and the shape of the specimen, the relation between the static (saturating) field and the microwave frequency, at resonance, is

$$\omega_0 = \gamma H$$

where γ is the gyromagnetic ratio (sections 2.1 and 5.11). In practice it is much easier to adjust d.c. field values than microwave frequencies; the frequency is held at a fixed value and the field is modulated to find the corresponding resonance value. According to the simplest approach, in which damping is neglected and the resonance field equated to that which gives the corresponding frequency of precession, there is no finite linewidth and the losses become infinite at a precise field value. Damping gives a finite linewidth, measured as the width, in oersteds, of the peak in μ'' (versus H) at half its height.

Structural imperfections, such as porosity and crystalline defects, and also the angular distribution of the easy directions, contribute to the linewidth. When these are largely eliminated by growing very good single crystals it is found that some spinel ferrites have low linewidths, of the order of 10 Oe, but the lowest linewidths are obtained for the ferrimagnetic garnets, particularly yttrium iron garnet $Y_3Fe_5O_{12}$. The Y^{3+} ions have no atomic moment, and the spontaneous magnetization arises from the unequal numbers of Fe^{3+} ions in each of two antiferromagnetically coupled sub-lattices. The exceptionally low linewidths which may be obtained for good crystals of yttrium iron garnet, down to $\Delta H = 0 \cdot 015$ Oe [11], are connected with the extremely regular ordering of the ions on the various types of site in the lattice. The ordering is far more regular than that in the spinel ferrites, in which it can usually only be said that most of the ions of one kind are on one particular kind of site. Disorder leads to the generation of spin waves, which can interchange energy with the lattice far more readily than can the uniform precessional mode.

For many microwave applications the value of the magnetic material is indicated by the narrowness of the resonance line, that is by the extent to which the losses are concentrated about one particular field value. Thus yttrium ion garnet, rare-earth iron garnets, or garnets of more complex composition are well suited for most microwave applications.

2.7.6 Fast switching

Many types of specimen give quite square hysteresis loops. Polycrystalline ferrite rings should have a remanence equal to about 90% of saturation if each grain remains saturated, a condition which is closely connected with the absence of demagnetizing fields for this particular geometry. Since the remanent magnetization can run in either of two senses around the ring, the ferrite cores can be used to store information on the binary system.

The information can be written in by sending a pulse of either positive or negative polarity along a single wire threading the core. The two alternative remanent states correspond to either a 'one' or a 'zero'. This information can subsequently be read out by sending a further pulse along the wire: if the first pulse was positive, storing a 'one' say, a second positive pulse will have little effect but a negative pulse of sufficient amplitude will cause the core to switch. The switching is indicated by the induction of a pulse in a further 'sense' line threading the core.

In coincident current stores the operation is a little more complicated, but the first specification, that the cores should have square loops, is the same. This property is shared by films of nickel–iron alloy (Permalloy), formed by evaporation in a vacuum onto a heated substrate in the presence of a static magnetic field. The field induces an anisotropy which is uniaxial in the plane of the film; otherwise the randomly-oriented polycrystalline film should be isotropic. The square loops measured parallel to this easy axis are due to the anisotropy combined with the small demagnetizing factors. The films are only about 10^{-4} mm thick and 1 mm across.

Both films and (very small) ferrite cores are used for memory stores in computers. The information must be capable of being handled very rapidly indeed, and the speed with which the elements can be switched is of extreme importance for this application. Thus, the dynamics of magnetization switching has great technical importance as well as being of inherent interest, and is dealt with fairly extensively (sections 2.1; 5.10.1; 6.7, and 6.8).

2.7.7 Further applications of magnetic materials and principles

Although the principal uses of magnetic materials are mentioned in the foregoing sections, the list is far from comprehensive. Transducers, for example, utilize materials with high magnetostriction. The construction of recorder heads calls for materials (usually ferrites, particularly for video recording at ultra high frequencies) with very good mechanical properties in order to facilitate the fine machining required.

Special reference might be made to works by C.D.Mee [12] for details of magnetic recording materials. The principles involved are very similar to those concerned in the sections on permanent magnets. Powders for recording tapes are basically permanent magnet materials (see Chapter 5) with relatively low coercivity, and gamma iron oxide (γ-Fe_2O_3) is used extensively. Data on recording materials are given in Table 2.2.

Finally it should be stressed that the importance of magnetic measurements is very great in the fields of both technical and theoretical physics and chemistry, and appears still to be increasing. The relevance of susceptibility or magnetization measurements to problems concerning atomic and molecular structures has long been appreciated. However, more recently the techniques of paramagnetic resonance, ferro- and ferrimagnetic resonance, nuclear magnetic resonance and various relaxation measurements have been applied to such problems with outstanding results. Probably the most remarkable development, in terms of the number of instruments in use and the sheer quantity of information obtained, has been the application of paramagnetic and nuclear magnetic resonance to the determination of ionic and molecular structure.

Table 2.2a. Physical and magnetic properties of oxide powders used for magnetic recording.

Material	Composition	Crystal structure	Density ρ	Length (μm)	Width (μm)	H_0	σ_s (gauss cm³/g)	I_s'' (gauss)	I_r'' (gauss)	H_{at}^* (Oe)	K_1 (erg/cm³)	H_a (Oe)	T_C (°C)
Iron oxide	γFe_2O_3	Inverse spinel	4·98	1·0	0·2	250	80	160	120§	2000	0·047	2500**	675
	Fe_3O_4		5·21	0·2	0·2	90	80	160	44‖		0·047	235††	675
Cobalt–iron oxide	$CoFe_2O_4$	spinel	5·29	0·2	0·2	115	92	192	52‖	245	0·11	460††	585
	$Co_xFe_{3-x}O_4$†		5·0	0·08	0·08	4200	80	170	120‡‡	1400	2·5	12000††	520
						640	73	146	100‡‡	1300	1·0	5500††	
Barium ferrite	$BaFe_{12}O_{19}$	Hexagonal close packed	5·3	1·0	0·1	4000	72	148	112§	17500	3·0	17000††	450
Barium-iron oxide + titanium–cobalt	$BaCo_\delta Ti_\delta Fe_{12-2\delta}O_{19}$‡		5·34	1·0	0·1	1900	60	128	96§		1·3	8200††	
Lead ferrite	$Pb_{1.68}Fe_{11.5}O_{19}$		5·5			500	49	110	82§				
Chromium oxide	CrO_2+Sb	Tetragonal	5·0	1·0	0·1	380	90	160				(100)††	130
Manganese ferrite	$MnFe_2O_4$	Inverse spinel					80						300
Nickel ferrite	$NiFe_2O_4$	Inverse spinel	5·33	0·06	0·06	123	49	105				412††	585

* H_{at} is the estimated anisotropy field at 150°C.
† x = 0·15.
‡ δ = 0·5.
§ Estimated value for partially oriented particles ($I_r''/I_s'' = 0.75$); assumed volume packing factor = 0·4.
‖ $I_r''/I_s'' = 0.27$.
** $H_a = 2\pi I_s$.
†† $H_a = 2K_1/I_s$.
‡‡ $I_r/I_s = 0.7$.

Table 2.2b. Physical and magnetic properties of some metals and alloys (in powder and film form) used for recording (C.D. Mee) [12].

Material	Composition	Crystal structure	Form	Density ρ'	Length a (μm)	Width, b (μm)	H_c (Oe)	σ_s $\left(\frac{gauss\,cm^2}{g}\right)$	I''_s (gauss)	I''_r (gauss)	H_{st} (Oe)	K_1 $\left(\frac{erg}{cm^3}\right)$	H_a (Oe)	T_C (°C)
Iron		b.c.c.	Powder	7·88	0·04	0·015	825	218	680	460	9800	0·4	10000‖	770
			Thin film	7·88	0·04	0·015	400	218	1700‡	1350‡				770
Cobalt		h.c.p.	Powder	8·9	0·04	0·015	2100††	157	560		2000	4·1	5500	1120
Nickel		f.c.c.	Powder	8·9				54	194		90	−0·05	206	358
Iron–cobalt	60% Fe, 40% Co	b.c.c.	Powder	8·1	0·1	0·02	1075	235	750	565†	11800	0·04	12000‖	1000**
Cobalt–nickel	82% Co, 18% Ni	h.c.p.	Thin film	8·9			250	143	1230	800				
Cobalt–phosphorous	98% Co, 2% P*	hex.	Thin film	8·9	2 μm layer		400	73	650	440				
Cobalt–nickel–phosphorous	75% Co, 23% Ni, 2% P*	hex.	Thin film	8·9			750	95	840	630				
Iron–cobalt–nickel	55% Fe, 5% Ni, 40% Co, 5% Ni	b.c.c.	Powder		0·1	0·05	760		314	239				
Iron–nickel–chromium	76% Fe, 12% Ni, 12% Cr	b.c.c.	Thin film				250		160	120				100
	74% Fe, 8% Ni, 18% Cr	f.c.c.	Thin film				200		240	240				
Vicalloy II (vanadium–iron–cobalt)	13% V, 35% Fe, 52% Co	b.c.c.	Thin film	8·2			500		1730§					
Cunife I (copper–nickel–iron)	60% Cu, 20% Ni, 20% Fe	f.c.c.	Thin film	8·1			590			450				
Cunife II	50% Cu, 20% Ni, 2·5% Co, 27·5% Fe	f.c.c.	Thin film	8·6			260			580§				
Cunico I (copper–nickel–cobalt)	50% Cu, 21% Ni, 29% Co	f.c.c.	Thin film	8·3			700			275				
Cunico II	35% Cu, 24% Ni, 41% Co	f.c.c.	Thin film	8·3			450			420				

* Estimated nominal phosphorous content.
† Estimated for partially oriented particles ($I''_r/I''_s = 0.75$);
‡ I''_r, I''_s are theoretical values.
§ Directional properties.
‖ $H_a = 2\pi I_s$.
** Virtual T_C.
†† Random orientation 76°K.

References

1. P. Weiss, *J. Phys.*, **6**, 667 (1906).
2. W. Pauli, *Z. Physik*, **41**, 81 (1927).
3. E. C. Stoner, *Proc. Roy. Soc.*, **A154**, 656 (1936).
4. D. J. Craik, *J. Sci. Instr.*, series 2, **1**, 1193 (1968).
5. S. Foner, *Rev. Sci. Instr.*, **27**, 548 (1956).
6. S. Foner, *Rev. Sci. Instr.*, **30**, 548 (1959).
7. R. M. Bozorth and H. J. Williams, *Phys. Rev.*, **103**, 572 (1956).
8. H. A. Mook and C. G. Shull, *J. Appl. Phys.*, **37**, 1034 (1966).
9. U. Enz, *Proc. Inst. Elec. Engrs. (London)*, **109B**, 246 (1961).
10. H. P. J. Wijn and H. Van der Heide, *Rev. Mod. Phys.*, **25**, 98 (1953).
11. E. G. Spencer and R. C. Le Craw, *Proc. Inst. Elec. Engrs. (London)*, **109B**, 66 (1961).
12. C. D. Mee, *"The Physics of Magnetic Recording"*, North Holland, 1963; *Inst. Elec. Electron. Engrs. Trans. Communications and Electronics*, 399 (1964).
13. W. E. Henry, *Phys. Rev.*, **88**, 559 (1952).
14. F. Tyler, *Phil. Mag.*, **11**, 596 (1931).
15. A. Vassiliev, J. Nicolas, and M. Hildebrandt, *Compt. rend.*, **252**, 2681 (1961).
16. E. W. Gorter, *Philips Res. Rept.*, **9**, 295 (1954).
17. C. G. Shull, W. A. Strauser, and E. O. Wollan, *Phys. Rev.*, **83**, 333 (1951).

3

Atomic Structure

The account of the magnetic behaviour of different classes of solids, given in the previous chapter, was remarkably successful in a qualitative way since it treated only collections of magnetic dipoles behaving in a classical manner. To make the account quantitative it is first necessary to calculate the values of the magnetic moments associated with atoms or ions, and this is a quantum mechanical problem. It will also be seen that some of the details of the classical calculations must be modified to bring them into line with quantum mechanical principles.

The study of atomic magnetic moments is virtually atomic theory itself, and must thus be introduced in a very general manner. It is impossible to be comprehensive in the present context, but it does seem important to start from first principles even if many results and rules must eventually be quoted.

3.1 BOHR THEORY—THE BOHR MAGNETON

The Bohr theory is very simple and limited in scope. It is included here because it makes the relation between angular momentum and magnetic moments readily apparent, and some of the results are carried over into more advanced quantum theory.

A one-electron atom is treated as a stationary, relatively massive, positively charged nucleus (charge $+e$) around which an electron revolves with velocity v in a circular orbit of radius r. The electron has mass m and charge $-e$, and thus has an angular momentum mvr. According to Chapter 1 there is an associated magnetic dipole

$$\mu = \text{current} \times \text{area enclosed} = \frac{ev}{2\pi r} \times \pi r^2 = \frac{evr}{2c} \text{ e.m.u.} \qquad (3.1)$$

where the factor c implies that the value of e is given in e.s.u.

The situation described is not stable. Classically the electron should radiate electromagnetic energy (motion in a circle implies acceleration even if the magnitude of $\mathbf{v}$ is constant) and since the velocity should then decrease there could never be a balance between the electrostatic and centrifugal forces and the orbit should collapse. To account for the stability Bohr introduced the restriction expressed by

$$mvr = n\frac{h}{2\pi} \qquad n = 1, 2, 3 \ldots \qquad (3.2)$$

This states that the angular momentum cannot vary continuously but can only have values which are integral numbers of a constant (Planck's constant divided by 2π). This remarkable postulate is acceptable because it leads to results which can be verified by spectroscopy. (A more meaningful approach may be as follows: the light emitted by an atom after subjection to a treatment which can be inferred to raise the energy of its electrons, that is an excitation process, is found to have a limited number of discrete frequencies. If we apply Planck's equation, $\epsilon = hf$, it follows that the atoms emit photons with discrete energies ϵ_1, ϵ_2, etc., which presumably originate from changes in the energy of the electrons between discrete levels. These levels can be identified by spectroscopy or measurements of the frequencies and it can then be shown that Equation (3.2) is the necessary restriction on angular momentum corresponding to the observed restriction to particular energy levels.)

On eliminating v from Equations (3.1) and (3.2) it becomes apparent that the magnetic moment must itself be quantized according to

$$\mu = \frac{neh}{4\pi mc} = n\mu_B \tag{3.3}$$

where μ_B is the quantum unit of magnetic moment, the Bohr magneton:

$$\mu_B = \frac{eh}{4\pi mc} . \tag{3.4}$$

It is also seen that the ratio of magnetic moment to angular momentum (the magneto-mechanical ratio) is

$$\frac{neh}{4\pi mc} \bigg/ \frac{nh}{2\pi} = \frac{e}{2mc} . \tag{3.5}$$

This expression is also carried over into further quantum theory, with one very important modification to be discussed later, which gives the magneto-mechanical ratio as

$$\gamma = \frac{ge}{2mc} . \tag{3.6}$$

Further developments of the Bohr theory are not very fruitful. It may be emphasised that the magnetic moments are vectors; for example, two superimposed orbits with opposite senses of rotation give zero resultant moment.

3.2 BASIS OF QUANTUM THEORY

Quantum theory is of universal application, but is only of real importance in the case of very small particles. A collection of large particles, such as billiard balls or the solar system, can be characterized by a series of measurements of the coordinates and velocities (or by the momenta $\mathbf{p} = m\mathbf{v}$ if uniform mass is not assumed) and it is assumed that such measurements can, in principle, be made simultaneously and with a precision limited only by the instruments used. The results of the measurements can be considered to give rise to the state function, $\psi(\mathbf{r}_1 \ldots \mathbf{r}_n, \mathbf{p}_1 \ldots \mathbf{p}_n)$ for n particles, which carries all the information about the system. The kinetic energy, E, is included implicitly since $E = \frac{1}{2}mv^2 = p^2/2m$, as is also the potential energy since it is a function of the coordinates.

The following assumptions are made about such so-called classical systems, and in fact define what is meant by 'classical' in this sense:

(a) Observations of position, velocity, etc. do not affect the system itself; they would simply be considered to involve poor experimental technique if they did so.

(b) From (a) it follows that, in principle, all the properties of the system can be determined simultaneously and precisely.

(c) It is immaterial in which order the observations are carried out.

In fact the above assumptions are approximations and are valid only when the particles comprising the system are large: any observation must involve *some* interaction with the system and hence cause an alteration of the property observed. It is traditional to deal with the limiting case of visual observation, or the interruption of a light beam, which must involve a collision with at least one photon with energy hf, or $\hbar\omega$ where $\hbar = h/2\pi$ and $\omega = 2\pi f$.[1] Moreover, when only small particles are involved, the frequency cannot be indefinitely small or the particle will not be resolved due to the long wavelength associated with the photon ($f = c/\lambda$). For particles the size of electrons such arguments suggest a disturbance of the order of Planck's constant h.

Consequently, the observations must correspond to the following two principles, of importance when they relate to very small particles:

(a) Observation A followed by observation B gives results which are different to those obtained when observation B is followed by A. A and B simply stand for any two different observations, for example position and momentum.

(b) To be meaningful the observations must still give rise to sets of numbers as their results.

Principle (a) suggests that the observations might be represented by operators of the differential type, because if A is represented by operator $\hat{A}$ and B by $\hat{B}$, then the statement is equivalent to

$$\hat{A}\hat{B} \neq \hat{B}\hat{A} \quad \text{or} \quad \hat{A}\hat{B} - \hat{B}\hat{A} \neq 0. \tag{3.7}$$

---

[1] It will be recalled that light can be represented either as a stream of particles with energy $E = hf$ and momentum $p = E/c$ derived from the relativistic equation $E^2/c^2 = p^2 + m^2c^2$ with the mass $m = 0$, or as a wave motion, that is as a disturbance which varies with time and position according to $A \exp[-i(\omega t - \mathbf{k} \cdot \mathbf{x})]$. $\omega = 2\pi f$ and $k = 2\pi/\lambda$ and is measured in the direction of the wave. Thus, in addition $\omega = kc$, and $\mathbf{p} = \hbar\mathbf{k}$. In the same way electrons, which have distinct particle properties, were shown by de Broglie to be also represented by a wave motion.

$\hat{A}\hat{B} - \hat{B}\hat{A}$ is called the *commutator* of $\hat{A}$ and $\hat{B}$, and Equation (3.7) states that A and B do not *commute*. The commutator is more briefly written as

$$[\hat{A}, \hat{B}] = \hat{A}\hat{B} - \hat{B}\hat{A}. \tag{3.8}$$

For example, if $\hat{A}$ is $\partial/\partial x$ and $\hat{B}$ is simply x the commutator can readily be found by applying the operators to any function of x, $f(x)$ as

$$\left(x\frac{\partial}{\partial x} - \frac{\partial}{\partial x}x \right) f(x) = x\frac{\partial}{\partial x}f(x) - \frac{\partial}{\partial x}xf(x)$$

$$= x\frac{\partial}{\partial x}f(x) - x\frac{\partial}{\partial x}f(x) - f(x)\frac{\partial}{\partial x}x$$

$$= -f(x)(1).$$

i.e.

$$\left[x, \frac{\partial}{\partial x} \right] f(x) = -f(x).$$

$$\left[x, \frac{\partial}{\partial x} \right] = -1.$$

Note that the only mathematics used is the differentiation of a product.

Principle (b) reinforces the idea of associating observables with operators, because it is known that when differential operators are applied to eigenfunctions they generate a series of eigenvalues. For example, since

$$\frac{\partial}{\partial x}(e^{ax}) = a(e^{ax})$$

$$\frac{\partial^2}{\partial x^2}(\sin ax) = -a^2(\sin ax),$$

we say that e^{ax} (or $\sin ax$) is an *eigenfunction* of $\partial/\partial x$ (or $\partial^2/\partial x^2$) and a (or $-a^2$) is the *eigenvalue*. Equations of the above form adequately define the two terms introduced. Thus, if the act of observation is represented by an operator, the result of the observation will be an eigenvalue, so long as the state function is (or contains) an eigenfunction of the particular operator.

The reader should note the great importance of the form of the operator itself. When this has been chosen, the state function for an electron follows, so long as it is assumed that the electron must have specific values of the property concerned.

The operators of most importance in magnetic studies are those for angular momentum. Altogether operators are required for

position	$\hat{x}$
linear momentum	$\hat{p}$
components of angular momentum	$\hat{m}$
total angular momentum	$\hat{M}$
energy	$\hat{H}$: the Hamiltonian operator

3.3 LINEAR MOMENTUM

The commutator of two operators indicates the extent to which the two related observations interfere with each other. Thus, if the commutator is found to be zero, the two observations are quite compatible and the values of the observables can be known simultaneously and precisely, as in a classical system.

In the case of position and momentum it is postulated that

$$[\hat{x}, \hat{p}] = \alpha\hbar \qquad (= \alpha h/2\pi)$$

where α is to be determined. We now state that the position operator may be simply represented by the position coordinate, x, and then derive a corresponding expression for p. It has been shown above that $[x, \partial/\partial x] = -1$, and repeating the demonstration with $-\alpha\hbar\partial/\partial x$ in place of $\partial/\partial x$ it is seen that

$$\left[x, -\alpha\hbar\frac{\partial}{\partial x}\right] = \alpha\hbar$$

and thus

$$\hat{p} = -\alpha\hbar\frac{\partial}{\partial x} \, ,$$

and the eigenfunction for linear momentum is $\exp(-px/\alpha\hbar)$ since

$$-\alpha\hbar\frac{\partial}{\partial x}\exp(-px/\alpha\hbar) = p\exp(-px/\alpha\hbar) \ .$$

However, the eigenfunction for an electron moving in one dimension has been found independently by de Broglie to have the form

$$\exp(-ikx) = \exp(-ipx/\hbar),$$

and thus consistency is obtained by putting $\alpha = i$, and

$$\hat{p} = -i\hbar\frac{\partial}{\partial x} \qquad \left(= +\frac{\hbar}{i}\frac{\partial}{\partial x}\right) \tag{3.9}$$

This is, in fact, the operator for the x-component of the linear momentum, and corresponding relations apply to the y and z components.

3.4 COMPONENT OF ANGULAR MOMENTUM

An electron's motion in a central potential may be represented by the polar coordinates (r, θ, ϕ) as shown in Figure 3.1. The coordinate involved in the z-component of the angular momentum is ϕ, and if we represent the operator for this component by $\hat{m}_z$ an equivalent argument to that of the previous section gives

$$[\hat{\phi}, \hat{m}_z] = i\hbar$$

$$\hat{m}_z = -i\hbar\frac{\partial}{\partial\phi}, \tag{3.10}$$

and the eigenfunctions are given by

$$\Phi = e^{im\phi}, \tag{3.11}$$

where m gives the possible values (eigenvalues) of the z-component of the angular momentum.

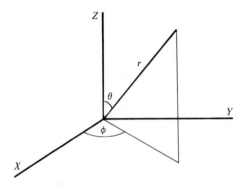

Figure 3.1. Coordinate system for the motion of an electron in a central potential.

In this case it is easy to find the numerical values of m, since the boundary conditions for rotation about a centre of motion are apparent, namely a rotation through ϕ equal to 2π restores the physical condition and $\Phi(\phi) = \Phi(\phi + 2\pi) = \Phi(\phi + 2n\pi)$ where n is any integer. This condition only applies to Equation (3.11) if m is an integer or zero, as can be seen by writing

$$e^{im\phi} = \cos m\phi + i \sin m\phi.$$

(Cosine and sine terms are periodic in 2π, but if m has the value $\frac{1}{4}$ for example, a change in ϕ of 2π would change $m\phi$ by $\pi/2$, which would not restore the values of $\sin m\phi$ and $\cos m\phi$.) Thus, the possible values of m are

$$m = 0, \pm 1, \pm 2, ... \tag{3.12}$$

and the eigenvalues are given by

$$-i\hbar \frac{\partial}{\partial \phi} e^{im\phi} = (+1)m\hbar e^{im\phi}.$$

Hence, the eigenvalues are $m\hbar$ or, with Equation (3.12):

$$m_z = m\hbar = 0, \pm\hbar, \pm 2\hbar, ... \tag{3.13}$$

This last equation is a result of real importance from the magnetic point of view. For an isolated atom there can, initially, be no possible real significance attached to any one arbitrary direction in space. If a uniform field is applied, however, it can be taken to define OZ and

Equations (3.13) and (3.5) indicate that the component of the magnetic moment associated with one electron, in this direction, is quantized and can only have the discrete values

$$\mu_z = m\hbar \times \frac{e}{2mc} = m\frac{eh}{4\pi mc} = m\mu_B \qquad (3.14)$$

that is

$$\mu_z = 0, \pm\mu_B, \pm2\mu_B \text{ etc.}$$

A common alternative symbol for the component of angular momentum $m\hbar$ is l_z or, since it is a vector, $\mathbf{l}_z$. m is sometimes replaced by m_l to stress the distinction from m_s (see section 3.8).

3.5 TOTAL ANGULAR MOMENTUM, $\hat{\mathbf{M}}$

From this point onwards the treatment is far from complete. The same principles are followed, with the addition of the correspondence principle, but the calculations become so lengthy that only the results will be indicated. The correspondence principle states that in the limit, when the system consists of particles so large that the disturbances corresponding to observation become negligible, the quantum mechanical operators must become the same functions as those which describe the properties in the classical manner.

The operator for the square of the total angular momentum, $\hat{M}^2$, is found from

$$\hat{M}^2 = \hat{m}_x^2 + \hat{m}_y^2 + \hat{m}_z^2,$$

after deriving $\hat{m}_x$ and $\hat{m}_y$, as

$$\hat{M}^2 = -\hbar^2\left[\frac{1}{\sin\theta}\frac{\partial}{\partial\theta}\left(\sin\theta\frac{\partial}{\partial\theta}\right) + \frac{1}{\sin^2\theta}\frac{\partial^2}{\partial\phi^2}\right], \qquad (3.15)$$

a function of θ and ϕ, but not of r.

The reader may be able to confirm that

$$[\hat{M}^2, \hat{m}_z] = 0 \qquad (3.16)$$

and hence it is possible to know both exactly; they are not subject to the uncertainty principle. This fact is important, because it means that there must be common eigenfunctions, functions of θ and ϕ, say $Y_{\beta m}(\theta, \phi)$, of such a form that they satisfy simultaneously both

$$\hat{m}_z Y_{\beta m} = \hbar m Y_{\beta m} \qquad (3.17)$$

and

$$\hat{M}^2 Y_{\beta m} = \hbar^2 \beta Y_{\beta m} \qquad (3.18)$$

where $\hbar^2\beta$ is the actual value of the square of the total angular momentum. This condition is satisfied if

(a) a factor in $Y_{\beta m}(\theta,\phi)$ is $e^{im\phi}$, and

(b) the remainder is a function of θ only, or

$$Y_{\beta m}(\theta,\phi) = P_{\beta m}(\theta)e^{im\phi}$$

such that

$$\hat{M}^2 P_{\beta m} = \hbar^2\beta P_{\beta m}. \qquad (3.19)$$

It is not surprising that $P_{\beta m}$ should depend on both the indices, β and m, since $\hat{M}$ contains $\hat{m}_z$ while the eigenfunctions for $\hat{m}_z$ depend only on the values of m.

The functions $P_{\beta m}$ are found to be standard functions (associated Legendre polynomials) generally specified as P_l^m where $l(l+1) = \beta$ and l must be integral. Low-index examples are given in Table 3.1, together with the eigenfunction for m.

Table 3.1. Normalized eigenfunctions for angular momentum.

l	$m = 1$	$m = 0$	$m = -1$
0		$Y_0^0 = \left(\dfrac{1}{4\pi}\right)^{\frac{1}{2}}$	
1	$Y_1^{+1} = -\left(\dfrac{3}{4\pi}\right)^{\frac{1}{2}}\sin\theta\,\dfrac{e^{i\phi}}{\sqrt{2}};$	$Y_1^0 = \left(\dfrac{3}{4\pi}\right)^{\frac{1}{2}}\cos\theta;$	$Y_1^{-1} = \left(\dfrac{3}{4\pi}\right)^{\frac{1}{2}}\sin\theta\,\dfrac{e^{-i\phi}}{\sqrt{2}}$

Another most important result which comes out of the full treatment is that l is always greater than $|m|$; in other words whatever value is chosen for l:

$$m = 0, \pm1, \pm2, ..., \pm l, \qquad (3.20)$$

which explains the gaps in the table. Thus m_z $(= m\hbar)$ has $(2l+1)$ values for each single value of l, corresponding to the familiar 'vector model'. Note also that the permitted values of the total angular momentum are

$$|\mathbf{l}| = \hbar[l(l+1)]^{\frac{1}{2}} \qquad (3.21)$$

and the magnitude of the magnetic moment for the single electron is

$$\mu = \frac{e}{2mc}\frac{h}{2\pi} = \mu_B[l(l+1)]^{\frac{1}{2}} \qquad (3.22)$$

The use of the same letter to indicate both the quantum number governing the angular momentum and the angular momentum itself may seem very confusing. However, the distinction is clearly made by remembering that the angular momentum is a vector, $\mathbf{l}$, while the quantum number is scalar, l. In turn this precludes the convention of representing the magnitude of the vector, in bold type, by its normal type symbol: $|\mathbf{l}| = l$ does *not* apply. It seems, in fact, to be acceptable in the literature to write simply

$$\mathbf{l} = \hbar[l(l+1)]^{\frac{1}{2}},$$

which is a very odd equation indeed unless one accepts that the right hand side stands for the magnitude of $\mathbf{l}$.

3.6 ENERGY, Ĥ

The operator for the energy must, by the correspondence principle, be the sum of the kinetic and potential energies, that is $\hat{H} = \hat{E} + \hat{V}$. $\hat{H}$ is known as the *Hamiltonian* operator, after the mathematician who applied the corresponding classical function to the generalized study of particle dynamics well before the conception of quantum mechanics.

For a particle moving in one dimension

$$\hat{E} = \frac{\hat{p}^2}{2m} = \frac{1}{2m}\left(-i\hbar\frac{\partial}{\partial x}\right)^2 = -\frac{\hbar^2}{2m}\frac{\partial^2}{\partial x^2},$$

since $\left(\dfrac{\partial}{\partial x}\right)^2 = \dfrac{\partial}{\partial x}\dfrac{\partial}{\partial x} = \dfrac{\partial^2}{\partial x^2}$. For an electron moving in a central potential, $V(r)$,

$$\hat{H} = \left[-\frac{\hbar^2}{2m}\nabla^2 + V(r)\right]. \qquad (3.23)$$

It is possible to show (after expressing $\hat{H}$ in polar coordinates) that

$$[\hat{H}, \hat{M}^2] = 0,$$

and

$$[\hat{H}, \hat{m}_z] = 0.$$

Thus, if the eigenfunction is Ψ_{nlm} it must factorise as

$$\Psi_{nlm}(r, \theta, \phi) = u_{nl}(r) Y_l^m(\theta, \phi).$$

The radial function u_{nl} is a function of r only, since the part of the Hamiltonian containing θ and ϕ is just the operator $\hat{M}^2$.

The index n is shown to be integral, and is known as the *principal quantum number*. Also $n > l$ so that

$$l = 0, 1, 2, \dots (n-1). \qquad (3.24)$$

It follows from the above that, for a one-electron atom (or a single electron in isolation from any others) the energy, given by

$$\hat{H}\Psi_{nlm}(r, \theta, \phi) = E_n \Psi_{nlm}(r, \theta, \phi), \qquad (3.25)$$

depends upon n alone (hence the single index in E_n), and not on l or m. However, different values of l and m designate different eigenfunctions and when a number of eigenfunctions give the same energy they are said to be *degenerate*. The degree of degeneracy is the number of states which give the same energy, and on taking Equations (3.20) and (3.24) together this is found to be just n^2.

$u_{nl}(r)$ is only slightly dependent on l, and can be taken approximately as $\exp(-zr/an)$ where a is the radius given by the simple Bohr theory. To complete the wavefunctions a normalization constant must be added, that is a constant so determined that it makes the integral of the square of the wave function over all space equal to unity.

3.7 MULTI-ELECTRON ATOMS

When the effects of electron interactions are accounted for, as well as the partial screening of the nuclear charge by inner electrons (that is those having lower values of n than the electron considered) it is found that the energies can depend quite strongly upon l as well as n. Thus, part of the degeneracy is lifted, but the energy is still independent of m for isolated atoms with no external influences, since the axis to which m refers is quite arbitrary. The mathematical problem of calculating the energies of the levels for different combinations of n and l is very difficult, but they can be evaluated by spectroscopy.

The important Pauli exclusion principle states that no more than two electrons (see below) can have the same set of quantum numbers, n, l and m. This situation leads to the formation of a series of 'shells' for the atomic electrons: for example for $n = 1$, $l = 0$, $m = 0$, and the first shell contains two electrons. When $n = 2$, $l = 0$, $m = 0$; or $n = 2$, $l = 1$ and $m = 1, 2$, or -1 and the second shell contains eight electrons. The periodic table is built up by feeding electrons into the lowest available level, but because of the effect of l on the energy it does not always follow that the inner shells, with the lower values of n, are completed before any electrons enter a higher shell. Probably the most important example, in relation to magnetic materials, is provided by the filling of the lower levels of the $n = 4$ shell before the higher levels of the $n = 3$ shell, to give the 3d transition series.

The rules for nomenclature for the states of the electrons are as follows:

(a) the number gives the value of n;

(b) the letter indicates the value of l by s, p, d, f, g, h which represent $l = 0, 1, 2, 3, 4, 5$ respectively. Thus, 3d electrons have $n = 3$, $l = 2$, $m = 2, 1, 0, -1, -2$, and the 3d subshell can accommodate ten electrons.

3.8 THE ZEEMAN EFFECT AND ELECTRON SPIN

If a magnetic field is applied to a free atom which has a magnetic moment the Hamiltonian must be modified, because the energy must contain a term due to the energy of orientation of the moment with the applied field. Classically this energy is $\boldsymbol{\mu} \cdot \mathbf{H} = \mu H \cos\theta$, that is the field × component of magnetic moment in field direction. However, this is $m\hbar(e/2mc)H$ (Equation 3.14), and the operator for m_z is $\hat{m}_z$, so that the total Hamiltonian in the presence of the field is

$$\hat{H}_0 + \frac{e}{2mc} H \hat{m}_z$$

and the energies are given by

$$\left\{ \hat{H}_0 + \frac{e}{2mc} H \hat{m}_z \right\} \psi_{nlm}(r,\theta,\phi) = \left\{ E_n + \frac{e}{2mc} H m \hbar \right\} \psi_{nlm}(r,\theta,\phi).$$

Since there are $2l+1$ different values of m for each value of l, there result $2l+1$ different values of the energy. Thus, the otherwise degenerate energy levels are split, by the presence of the magnetic field, into levels with spacing

$$\Delta E = \frac{e\hbar H}{2mc} = H \frac{eh}{4\pi mc} = H\mu_B. \qquad (3.26)$$

Such *Zeeman splitting* is readily observed spectroscopically, but additional anomalous splitting is also found, even when $l = 0$. Hence there must be some additional source of angular momentum and magnetic moment which is not associated with orbital motion; it has instead become known as *electron spin* [1].

The exclusion principle, as stated in the previous section, suggests that the spin be represented by a quantum number which can have just two values. These are represented by s and are found to be $s = \pm\frac{1}{2}$. Thus the principle can be stated alternatively as the permitting of no more than *one* electron to be designated by the same values of the four quantum numbers n, l, m, and s. For example, 10 of the electrons of sodium form the first two closed shells (2 for $n = 1$, 8 for $n = 2$) and the eleventh has $n = 3$, $l = 0$ (3s) in the lowest energy, or ground, state. This is an orbital singlet and should not be split by a field. However, the results of spectroscopic examination (observation of the frequencies of light emitted by sodium vapour, and deriving from these the differences in the energy levels of the atoms by $h\nu = E_1 - E_2$) clearly show that the ground state is split by a field as shown by Figure 3.2. When $l \neq 0$ the 'anomaly' appears as an extra splitting, giving more levels than those predicted for the different values of m.

Despite the fact that s is known to have half-integral values the magnitude of the anomalous Zeeman splitting shows that, for a single electron, the associated magnetic moment is still very nearly one Bohr magneton and further that this can only be oriented either parallel or antiparallel to an applied field. This orientational restriction can be specified by a quantum number for the component of the spin which can have the values $m_s = \pm\frac{1}{2}$ only. Further, the relation between the spin angular momentum and the spin magnetic moment must include a factor, the spectroscopic splitting factor $g \doteq 2$, and the magnitude of the moment is

$$\mu = g\mu_B[s(s+1)]^{1/2}$$

with a component in the field direction

$$g\mu_B m_s = \pm\tfrac{1}{2}g\mu_B \doteq \pm\mu_B .$$

For generality the ratio of angular momentum to magnetic moment may be always represented by

$$\gamma = \frac{ge}{2mc}$$

with g equal to 1 for the orbital contribution. γ is the *magneto-mechanical ratio*, sometimes confused with the *gyromagnetic ratio* which is, in fact, the reciprocal of γ. Taking the precise value of $g = 2\cdot0023$ for a free electron spin, $\gamma = 1\cdot7609 \times 10^7\,s^{-1}\,Oe^{-1}$.

The way in which the presence of the electron spin can be indicated, by introducing a function of 'spin coordinates' as a factor in the wave-function, is described in Chapter 4. This function has the form of a matrix, and the corresponding spin operator, which must be applied to it to give the spin eigenvalues, $\pm\tfrac{1}{2}$, is also in matrix form.

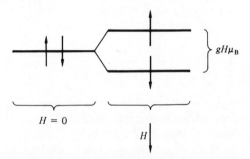

Figure 3.2. When $l = 0$, $m = 0$, $s = \tfrac{1}{2}$, and there are no external perturbing influences, the states with $m_s = \pm\tfrac{1}{2}$ have the same energy. An applied field lifts this degeneracy since one of the m_s values represents a spin magnetic moment parallel to the field (low energy) and vice versa.

3.9 COUPLING OF SPIN AND ORBITAL MOMENTA FOR MANY ELECTRONS

The mutual orientation of the spin momenta, and of the orbital momenta, for all the electrons in an atom, is not random but can be considered to be controlled by Hund's rules. These are empirically determined as the results of spectroscopy and will serve the present purpose, although they must obviously also follow from the basic concepts of quantum mechanics. The coupling forces are basically of an electrostatic nature, and involve an exchange energy as described in Chapter 4, but are certainly much stronger than any conceivable classical magnetostatic forces.

In a minority of cases (in heavy atoms, which are not of immediate interest) there is a direct interaction between the spin and orbital momenta of each electron. The momenta themselves (not the quantum numbers l and s) add vectorially to give a total momentum $\mathbf{j} = \mathbf{l} + \mathbf{s}$, which can itself be represented as $\hbar[j(j+1)]^{1/2}$.

For the 3d transition and rare-earth elements, as well as for those of lower atomic weight, the mutual coupling of the spins and the mutual coupling of the orbital momenta is stronger than the above. Thus, one obtains a single orbital momentum $\mathbf{L} = \Sigma\mathbf{l}$ and a single spin momentum $\mathbf{S} = \Sigma\mathbf{s}$ for the atom as a whole, which can be represented by the atomic quantum numbers L and S; $\mathbf{L} = \hbar[L(L+1)]^{1/2}$, $\mathbf{S} = \hbar[s(s+1)]^{1/2}$.

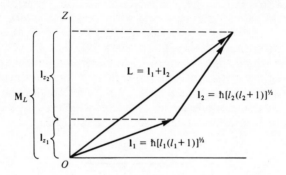

Figure 3.3. The vector addition of two angular momenta, and the effective scalar addition of their components. Since $\mathbf{M}_L = \hbar M_L = \hbar(m_{l_1} + m_{l_2})$, $M_L = m_{l_1} + m_{l_2}$ but $L \neq l_1 + l_2$ (where L is given by $\mathbf{L} = \hbar[L(L+1)]^{1/2}$.

Now if two vectors are added (in the vector sense) their components along any axis can be added in a scalar sense as illustrated by Figure 3.3. Thus the components of the angular momenta can be summed simply, with due regard to sign, to produce a total component of magnitude $\hbar M_L = \hbar\Sigma m_l$ and hence a total atomic quantum number $M_L = \Sigma m_l$. Note, however, that $L \neq \Sigma l$, an error which can be found quite frequently. It may be accepted as an obvious generalization of the single-electron case that M_L can have the $(2L+1)$ values

$$M_L = L, L-1, L-2, ..., -(L-1), -L;$$

(clearly the component cannot exceed the magnitude). Similarly, there are $(2S+1)$ values of $M_S = \Sigma m_s$, which can be either integral or half-integral.

$$M_S = S, S-1, S-2, ..., -(S-1), -S.$$

The coupling so far discussed, which gives rise to the separate atomic quantum numbers for spin and orbital momenta, suggests the representation of an atom by a 'spectroscopic term' designated by particular values of S and L. This is conveniently expressed by a number and a letter, symbolically ^{2S+1}L, putting in the actual value of S and the letter S, P, D, F, G or H for $L = 0, 1, 2, 3, 4$ or 5. $(2S+1)$ is used rather than S because (so long as $L \geqslant S$) this represents the multiplicity, or number of spin-degenerate states which have the same energy in the absence of any external influences such as applied magnetic fields. $S = 0$ gives a singlet, $S = \frac{1}{2}$ a doublet ($M_S = \frac{1}{2}$ or $-\frac{1}{2}$), $S = 1$ a triplet ($M_S = 1, 0, -1$), etc., as indicated by Figure 3.4.

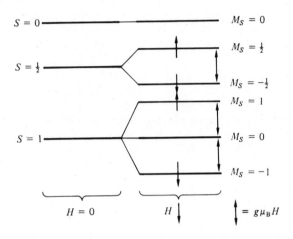

Figure 3.4. If $S = 0$ then $M_S = 0$ and there can be no spin degeneracy (singlet). The degeneracy of the doublet state: $S = \frac{1}{2}$, $M_S = \pm\frac{1}{2}$, and of the triplet state: $S = 1$, $M_S = +1$, $0, -1$, is illustrated by the splitting of the degenerate levels in an applied field.

We now return to the manner in which the coupling occurs in practice, that is, according to the preceding paragraph, to the question of which of the several terms that may be written down for a particular atom represents the lowest energy, or is the ground state. Hund's first two rules are as follows:

1. The values of m_s of the electrons in a subshell (all having the same value of n and of l) sum to give the maximum possible value of M_S consistent with the Pauli principle. Thus, for the 3d subshell the first five electrons have 'parallel spins', since there are five different values of m_l for $l = 2$, giving $S = M_S = \frac{1}{2}$ to $\frac{5}{2}$ progressively.

2. The values of m_l also sum to give the maximum possible M_L which is consistent with the Pauli principle and with rule 1.

The most apparent consequence of these rules is that, since a subshell by definition contains just that number of electrons which take up all the different combinations of m_l and m_s corresponding to the given l, we must have $M_L = M_S = L = S = 0$ for the full subshell. Thus, full subshells make no contribution to the angular momentum or the atomic magnetic moments (apart from the negative diamagnetic effect), and virtually disappear from our consideration.

Further development would be simpler if, for example, the iron transition series consisted of ten atoms which differed only by having additional 3d electrons. The situation is complicated by the presence of 4s electrons; that is by the failure of the 3d shell to fill up regularly from atom to atom in the series of ascending atomic numbers due to the existence of lower-lying 4s energy levels. However, the 4s electrons are those which are lost on ionization because of their spatial location and a regular 3d series is obtained for the divalent ions. The application of Hund's rules is then very simple, as it is for the trivalent ions of the palladium group and the rare earths, as shown by Table 3.2 and Figure 3.5. The first five (or seven) electrons have parallel spins which add directly to give S up to $\frac{5}{2}$ (or $\frac{7}{2}$). The maximum M_L for one electron is just the maximum m_l which is equal to l, and is thus 2 (or 3 for the 4f electrons). For the 3d case the second electron, having the same m_s as the first, must have a different m_l and so the maximum $M_L = 2+1 = 3$. Similarly, the third electron must have $m_l = 0$ while for the fourth $m_l = -1$ and L actually decreases. After the fifth electron all the combinations of $m_s = \frac{1}{2}$ with different values of m_l have been used, and the

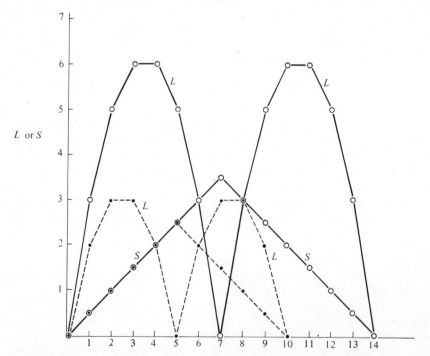

Figure 3.5. The total orbital angular momentum quantum number L, and the total spin quantum number S, for 3d or 4f shells containing the numbers of electrons shown, in conformity with Hund's rules. ○ for 4f and ● for 3d. Note $L = 0$ and $S = 0$ for full shells.

sixth electron must have $m_s = -\frac{1}{2}$ (while the Pauli principle then permits the use of $m_l = +2$ again) and the spin-pairing reduces the value of S once more. The remainder of Table 3.2 and Figure 3.5 follow the same principles.

3.10 SPIN-ORBIT COUPLING

The above explanation only accounts for a part of the coupling of the single-electron angular momenta, and thus their magnetic moments. The motion of the electrons in their orbits generates magnetic fields which interact with the spin magnetic moments and give rise to an energy term, W_{so} dependent on the mutual orientation

$$W_{so} = \lambda \mathbf{L} \cdot \mathbf{S}. \tag{3.27}$$

λ is the spin-orbit coefficient, the magnitude and sign of which can be determined experimentally by spectroscopy. It is found that λ is positive if the transition shell is less than half full, and negative if the shell is more than half full. This can be taken as a statement of Hund's third rule, since, for positive λ, W_{so} is a minimum when $\mathbf{L}$ and $\mathbf{S}$ are anti-parallel, that is

$$\mathbf{J} = \mathbf{L} - \mathbf{S}$$

for a shell less than half full, where $\mathbf{J}$ represents the total angular momentum. Similarly,

$$\mathbf{J} = \mathbf{L} + \mathbf{S}$$

for a shell more than half full.

Although it really calls for a lengthy discussion it may be accepted that the total angular momentum is related to a quantum number J and that the eigenvalues are

$$\mathbf{J} = \hbar[J(J+1)]^{\frac{1}{2}}.$$

The components of $\mathbf{J}$ are given by a similar relation to those for $\mathbf{L}$ and $\mathbf{S}$,

$$J_z = M_J \hbar$$

where M_J has the $(2J+1)$ values

$$M_J = J, J-1, J-2, ..., -(J-2), -(J-1), -J.$$

Also it can be shown that the operators for $\mathbf{J}$, J_z, $\mathbf{L}$ and $\mathbf{S}$ commute with each other and with the Hamiltonian, but not with those for M_L and M_S. Thus the state of an isolated atom or ion can be represented by the quantum numbers S, L and J, using the convention already noted and adding the value of J as a subscript. Due to the way in which $\mathbf{L}$ and $\mathbf{S}$ are ordered by the spin-orbit coupling, that is with parallel or anti-parallel orientation, M_J is simply the sum or difference of M_L and M_S. Taking L as the maximum value of M_L and S as the maximum value of M_S one obtains the maximum value of $M_J = J = L+S$ or $L-S$

Table 3.2. The configurations, atomic quantum numbers, ground terms and predicted ionic magnetic moments for isolated atoms or ions of (A) the iron group transition metals, (B) the palladium group, (C) the rare earths, and the ionic moments in units of Bohr magnetons p_e as derived from measurements on ionic salts.

A—Iron group, 3d; doubly ionized and neutral atoms.

Ion (atom)	Configuration	S	L	J	Ground term	$g[J(J+1)]^{1/2}$	p_e
Ca	- -	0	0	0	^1S		
Ca^{2+}	$4s^2$	0	0	0	$^1S_0$	0	(dia)
Sc	$3d4s^2$	$\frac{1}{2}$	2	$\frac{3}{2}$	^2D		
Sc^{2+}	$3d$	$\frac{1}{2}$	2	$\frac{3}{2}$	$^2D_{3/2}$	1·55	1·7
Ti	$3d^24s^2$	1	3	2	^3F		
Ti^{2+} (V^{3+})	$3d^2$	1	3	2	$^3F_2$	1·63	2·8
V	$3d^34s^2$	$\frac{3}{2}$	3	$\frac{3}{2}$	^4F		
V^{2+} (Mn^{4+})	$3d^3$	$\frac{3}{2}$	3	$\frac{3}{2}$	$^4F_{3/2}$	0·77	3·8
Cr	$3d^54s$	2	0		^5S		
Cr^{2+} (Mn^{3+})	$3d^4$	2	2	0	$^5D_0$	0·0	4·9
Mn	$3d^54s^2$	$\frac{5}{2}$	0	$\frac{5}{2}$	^6S		
Mn^{2+} (Fe^{3+})	$3d^5$	$\frac{5}{2}$	0	$\frac{5}{2}$	$^6S_{3/2}$	5·92	5·9
Fe	$3d^64s^2$	2	2	4	^5D		
Fe^{2+} (Co^{3+})	$3d^6$	2	2	4	$^5D_4$	6·70	5·4
Co	$3d^74s^2$	$\frac{3}{2}$	3	$\frac{9}{2}$	^4F		
Co^{2+} (Ni^{3+})	$3d^7$	$\frac{3}{2}$	3	$\frac{9}{2}$	$^4F_{3/2}$	6·64	4·8
Ni	$3d^94s$	1	2		^3D		
Ni^{2+}	$3d^8$	1	3	4	$^3F_4$	5·59	3·2
Cu	$3d^{10}4s$	$\frac{1}{2}$	0		^2S		
Cu^{2+}	$3d^9$	$\frac{1}{2}$	2	$\frac{5}{2}$	$^2D_{5/2}$	3·55	1·9
Zn	$3d^{10}4s^2$	0	0		^1S		
Zn^{2+} (Cu^+)	$3d^{10}$	0	0		$^1S_0$	0·0	(dia)

B—Palladium group, 4d; triply ionized.

Ion (atom)	Configuration	S	L	J	Ground term	$g[J(J+1)]^{1/2}$	p_e
Y^{3+}	-	0	0	0	$^1S_0$	0	
Zr^{3+}	$4d$	$\frac{1}{2}$	2	$\frac{3}{2}$	$^2D_{3/2}$	1·55	
Nb^{3+}	$4d^2$	1	3	2	$^3F_2$	1·63	0·7
Mo^{3+}	$4d^3$	$\frac{3}{2}$	3	$\frac{3}{2}$	$^4F_{3/2}$	0·77	3·6
Tc	-						
Ru^{3+}	$4d^5$	$\frac{5}{2}$	0	$\frac{5}{2}$	$^6S_{5/2}$	5·92	2·1
Rh^{3+}	$4d^6$	2	2	4	$^5D_4$	6·70	0·06
Pd^{3+}	$4d^7$	$\frac{3}{2}$	3	$\frac{9}{2}$	$^4F_{9/2}$	6·64	0·1
Ag^{3+}	$4d^8$	1	3	4	$^3F_4$	5·59	
Cd^{3+}	$4d^9$	$\frac{1}{2}$	2	$\frac{5}{2}$	$^2D_{5/2}$	3·55	

Table 3.2 continued.

C—Rare earth series, 4f: triply ionized.

Ion (atom)	Configuration	S	L	J	Ground term	$g[J(J+1)]^{1/2}$	p_e
La³⁺	$5s^2 5p^6$	0	0	0	$^1S_0$	0·00	(dia)
Ce³⁺ (Pr⁴⁺)	$4f^1 5s^2 5p^6$	$\frac{1}{2}$	3	$\frac{5}{2}$	$^2F_{5/2}$	2·54	2·4
Pr³⁺	$4f^2-$	1	5	4	$^3H_4$	3·58	3·6
Nd³⁺	$4f^3-$	$\frac{3}{2}$	6	$\frac{9}{2}$	$^4I_{9/2}$	3·62	3·8
Pm³⁺	$4f^4-$	2	6	4	$^5I_4$	2·68	
Sm³⁺	$4f^5-$	$\frac{5}{2}$	5	$\frac{5}{2}$	$^6H_{5/2}$	0·84	1·5
Eu³⁺ (Sm²⁺)	$4f^6-$	3	3	0	$^7F_0$	0·00	3·6
Gd³⁺ (Eu²⁺)	$4f^7-$	$\frac{7}{2}$	0	$\frac{7}{2}$	$^8S_{7/2}$	7·94	7·9
Tb³⁺	$4f^8-$	3	3	6	$^7F_6$	9·72	9·6
Dy³⁺	$4f^9-$	$\frac{5}{2}$	5	$\frac{15}{2}$	$^6H_{15/2}$	10·63	10·6
Ho³⁺	$4f^{10}-$	2	6	8	$^5I_8$	10·60	10·4
Er³⁺	$4f^{11}-$	$\frac{3}{2}$	6	$\frac{15}{2}$	$^4I_{15/2}$	9·59	9·4
Tm³⁺	$4f^{12}-$	1	5	6	$^3H_6$	7·57	7·3
Yb³⁺	$4f^{13}-$	$\frac{1}{2}$	3	$\frac{7}{2}$	$^2F_{7/2}$	4·54	4·5
Lu³⁺ (Yb²⁺)	$4f^{14}-$	0	0	0	$^1S_0$	0·00	(dia)

according to whether the spin-orbit coupling constant λ is negative or positive. Hence the subscripts in Table 3.2 are obtained simply by adding or subtracting the given values of S and L according to the state of occupation of the shells.

3.11 IONS IN APPLIED MAGNETIC FIELDS

To analyse the effect of magnetic fields on isolated ions it is first necessary to determine the magnitude of the ionic magnetic moment, and of its permitted components in the field direction. The same values of g (the Landé spectroscopic factor) apply to L and to S as apply to l and to s since, for example,

$$\mu \text{ (electron)} = g\mu_B s,$$

and

$$\mu \text{ (ion)} = \sum g\mu_B s = g\mu_B \sum s = g\mu_B S,$$

but the value of g which applies to $\mathbf{J}$ has to be determined, using $g = 1$ for the orbital contributions and $g = 2$ for the spin contributions.

The vector sum, $\mathbf{L} + \mathbf{S} = \mathbf{J}$, can be represented diagrammatically as in Figure 3.6. It must be noted, however, that states with definite values of M_J do not have definite values of M_L or M_S, and to make the diagram consistent with this it must be assumed that only the components of $\mathbf{L}$ and $\mathbf{S}$ parallel to $\mathbf{J}$ are fixed and otherwise they can have any orientation on the cones defined by the constant angles θ and ϕ shown. Since the associated magnetic moment vectors are themselves parallel to **S** and to **L**, the components of these normal to **J** lie in any direction around **J** and sum to zero.

The magnetic moment to be associated with **J** thus consists only of the components of the moments associated with **L** and **S** that are parallel to **J**. With an obvious notation

$$\mu_J = \frac{\mu_L \cdot \mathbf{J}}{|\mathbf{J}|} + \frac{\mu_S \cdot \mathbf{J}}{|\mathbf{J}|} \ .$$

(Note that μ_J, being defined in direction, is not itself a vector.)

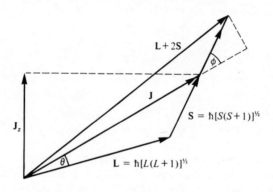

Figure 3.6. Illustration of the vector addition **J** = **L**+**S**. Note that **L**+2**S** is not parallel to **J**. θ and ϕ are constant but the directions of **L** and **S** are not otherwise defined; they are supposed to be instantaneously in the same plane for convenience.

Since $\mu_L = g\mu_B [L(L+1)]^{1/2}$ then $\mu_L = \dfrac{\mu_B}{\hbar} \mathbf{L}$ $(g = 1)$, and also $\mu_S = \dfrac{2\mu_B}{\hbar} \mathbf{S}$ $(g = 2)$, giving

$$\mu_J = \frac{\mu_B}{\hbar|\mathbf{J}|} [\mathbf{L} \cdot \mathbf{J} + 2\mathbf{S} \cdot \mathbf{J}]$$

$$= \frac{\mu_B}{\hbar|\mathbf{J}|} [\mathbf{L} \cdot (\mathbf{L}+\mathbf{S}) + 2\mathbf{S} \cdot (\mathbf{L}+\mathbf{S})]$$

$$= \frac{\mu_B}{\hbar|\mathbf{J}|} [L^2 + 2S^2 + 3\mathbf{L} \cdot \mathbf{S}]$$

(recalling that **L** · **S** = **S** · **L**, although **L** × **S** = −**S** × **L**). But since

$$J^2 = (\mathbf{L}+\mathbf{S})^2 = L^2 + S^2 + 2\mathbf{L} \cdot \mathbf{S}$$

then

$$3\mathbf{L} \cdot \mathbf{S} = \tfrac{3}{2}(J^2 - L^2 - S^2),$$

and

$$\mu_J = \frac{\mu_B}{\hbar|\mathbf{J}|}[\tfrac{3}{2}\mathbf{J}^2 - \tfrac{1}{2}\mathbf{L}^2 + \tfrac{1}{2}\mathbf{S}^2]$$

$$= \frac{\mu_B}{\hbar^2[J(J+1)]^{1/2}}[\tfrac{3}{2}\hbar^2 J(J+1) + \tfrac{1}{2}\hbar^2 S(S+1) - \tfrac{1}{2}\hbar^2 L(L+1)];$$

that is

$$\mu_J = \mu_B\left[\frac{3J(J+1) + S(S+1) - L(L+1)}{2J(J+1)}\right][J(J+1)]^{1/2},$$

which may be written

$$\mu_J = g\mu_B[J(J+1)]^{1/2},$$

where g has the value indicated. (g clearly satisfies the requirement that it reduces to 2 when $L = 0$, and to unity when $S = 0$.) It is necessary to make a careful distinction between μ_J as defined, for which the equation

$$\mu_J = \frac{g\mu_B}{\hbar}\mathbf{J}$$

is consistent since μ_J is parallel to $\mathbf{J}$, and the vector sum of μ_L and $2\mu_S$, which is not parallel to $\mathbf{J}$ (Figure 3.6).

Because μ_J is parallel to $\mathbf{J}$ and μ_z is parallel to $\mathbf{J}_z$, it is apparent by geometry that the same g values apply, that is g has the above value in

$$\mu_z = \frac{g\mu_B}{\hbar}\mathbf{J}_z, \qquad\qquad \mu_z = g\mu_B M_J.$$

Thus a field H applied in the z direction gives rise to $(2J+1)$ equally spaced discrete energy levels for each otherwise degenerate J state. The spacing is $g\mu_B H(\Delta M_J) = g\mu_B H$, as represented in Figure 3.7 for $J = 2$ and $J = \tfrac{3}{2}$. Putting $g \doteq 1$ and $H \doteq 10000$ Oe the magnitude of the splitting is of the order of $1\,\mathrm{cm}^{-1}$, which gives $\Delta E \doteq kT/200$ at room temperature.

The quantum mechanical treatment of susceptibility deals with the equilibrium distribution of the electrons between the energy levels as split by the applied field. When the splitting is so small compared to kT the distribution is nearly random, with only a slight excess population of the lower levels of the multiplet (those corresponding to a component parallel to the field) and the relative magnetization and susceptibility are small. The theory given in Chapter 2 is still applicable (using the summation over the possible states rather than integration for continuously varying orientation) and the inequality corresponds to the low-field, high-temperature approximation, giving

$$\chi = \frac{ng^2\mu_B^2 J(J+1)}{3kT}; \tag{3.28}$$

thus it is only necessary to replace the arbitrary moment μ by μ_J. The full expression for the magnetization induced in very high fields, at low temperatures, can be shown to be

$$I = ngJ\mu_B \left[\frac{2J+1}{J} \coth\left(\frac{2J+1}{2J} y_J \right) - \frac{1}{2J} \coth\left(\frac{1}{2J} y_J \right) \right] \qquad (3.29)$$

where

$$y_J = \frac{gJ\mu_B H}{kT} .$$

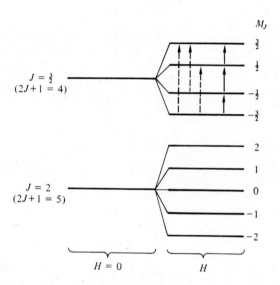

Figure 3.7. Zeeman splitting for J states (cf. Figure 3.4) due to a constant magnetic field. The full arrows indicate transitions which are found to occur in practice, the broken arrows represent 'forbidden transitions'.

The simplified expression represents the Curie law and when this law is found experimentally, susceptibility measurements can be used to give a measure of μ_e ('μ-effective'), or p_e (equal to μ_e/μ_B) the effective number of Bohr magnetons. These measured values can then be compared with the calculated values of $g[J(J+1)]^{1/2}$, or with the further developments.

Measurements on monatomic gases would be the best test of the theory for free atoms or ions. However, few of these have magnetic moments, apart from metallic vapours on which measurements are difficult: for potassium vapour [2] a value of p_e^2 equal to $3\cdot04 \pm 10\%$ has been obtained, in good agreement with expectation [ground state ^2S, $J = S = \frac{1}{2}$, $g = 2$, $g^2J(J+1) = 3$].

Measured values of p_e for ions of transition metals and rare earths in ionic crystals are included in Table 3.2. Discussion of the transition metals is reserved for the following chapter—the values of p_e given are certainly not explicable by the above principles. Rather surprisingly, however, for the rare earths it is seen that there is generally very good agreement between p_e and $g[J(J+1)]^{1/2}$; in other words these do behave as if they were free ions.

If we anticipate the following chapter, the basis for the difference between the two groups is associated with the outer electrons. It should not, in fact, be expected that the 'magnetic' electrons will be unaffected by the very strong electrostatic fields arising from neighbouring ions in the crystal lattice (that is the crystal fields or ligand fields) but the full 5s and 5p shells very effectively shield the 4f electrons in the rare earth ions.

The only important discrepancies between p_e and $g[J(J+1)]^{1/2}$ in the table are for Sm^{3+}, and more particularly Eu^{3+}. This is explicable, not in terms of crystal field effects directly, but because the identification of p_e with $g[J(J+1)]^{1/2}$ involves the assumption, implicit in the calculation of the susceptibility, that only the ground state was occupied. The distribution of electrons considered was between different M_J values, all of which were associated with the same J. However, it is found that for Eu^{3+}, the states with $J = 0$, $J = 1$, $J = 2$ and $J = 3$ all fall within $10kT$ at room temperature and thus the thermal distribution of the electrons between those different states should be taken into account. The electrons with a given value of J still contribute to the susceptibility as predicted above, but the distribution over different J values alters the temperature dependence giving:

$$\chi = n \frac{\sum\limits_{J=L-S}^{J=L+S} \{[g^2\mu_B^2 J(J+1)/3kT] + \alpha(J)\}(2J+1)e^{-E(J)/kT}}{\sum (2J+1)e^{-E(J)/kT}} . \quad (3.30)$$

In this expression $E(J)$ is the energy of the state J and the term $\alpha(J)$ can be considered as indicating an approximation in the vector model used above, that is the neglect of those components of the resolved magnetic moments associated with S and L which are normal to J.

3.12 PARAMAGNETIC RESONANCE

The relevance of susceptibility measurements to the general theory of atomic magnetism has already been explained, whilst that of spectroscopic measurements generally can be assumed. The importance of paramagnetic resonance should also be indicated briefly, although the current interest is mainly with magnetically ordered materials.

When a magnetic field is applied so that the electron energy levels are split as shown in Figure 3.7, the lower levels will develop a marginally higher population (which is just another way of saying that a certain level of magnetization is induced). In accordance with the principles of spectroscopy it should be possible to induce transitions to higher levels provided the energy fed into the system is in the form of quanta of the correct magnitude so that resonant absorption occurs, that is

$$\hbar\omega = g\mu_B H \,,$$

$$\omega = g\frac{2\pi}{h}\frac{eh}{4\pi mc}H = g\frac{e}{2mc}H = \gamma H \,, \qquad (3.31)$$

where γ is the magneto-mechanical ratio. This expression, however, is just the relation for the precessional frequency of a magnetic dipole as derived in a classical manner at the beginning of Chapter 2, and so there is a direct analogy between the simple quantum-mechanical principle of resonant absorption and the interaction of an oscillating field of the correct frequency with a precessing dipole. This latter approach is developed further in Chapter 5, in connection with ferromagnetic resonance: it shows, for example, why the oscillating field should be normal to the static applied field, H.

In writing down Equation (3.31) the existence of selection rules was assumed. The only transitions which are possible, according to the full quantum theory and to actual observations, are those for which $\Delta M_J = 1$, as indicated in Figure 3.7. Thus, for example, $\hbar\omega \neq 3g\mu_B H$ for absorption.

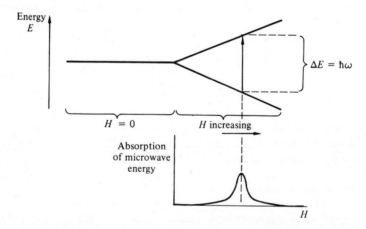

Figure 3.8. Schematic representation of the splitting of a doublet energy level by an applied field which increases in magnitude from left to right. Considering ω to be fixed resonant absorption occurs when $\Delta E = \hbar\omega$ as shown (cf. Figure 3.7).

Inserting numerical values in Equation (3.31), with $g = 1$, gives

$$f(\text{MHz}) = 1\cdot4H\,(\text{Oe}).$$

and with $H = 10^4$ the resonant frequency is in the microwave region, and waveguides and cavities must be used. In principle both H and f can be reduced, for technical convenience, but when this is done the resonance may have a linewidth which is large compared to the applied field and thus be quite impossible to detect. Accepting the use of microwaves it is easier to vary the field than the frequency, which gives a situation represented schematically by Figure 3.8 (as compared with Figure 3.7 in which it is assumed that H is constant). The simplest theory predicts absorption at a single frequency giving a true absorption line, but this 'line' is broadened to a peak by the realxation phenomena, which must exist for the resonance to be observed continuously.

Assuming that the linewidth for the material studied is not too large, the position of the line gives the value of g directly, and the great significance of this parameter should be apparent. Valuable information can also be derived from the fine details of the spectra and from the linewidths themselves.

References

1. P. A. M. Dirac, *Proc. Roy. Soc.,* **A117**, 610 (1928).
2. E. Lehrer, *Ann. Phys. (Leipzig),* **81**, 229 (1927).

4

Crystal Structure

In the study of magnetism generally, and particularly in the study of magnetic materials, the principal interest is with the magnetic properties of ions which form part of a crystal lattice. The magnetic properties of free ions are modified in two ways when they are incorporated into a crystal lattice.

Firstly the magnitudes of the ionic magnetic moments are strongly influenced by the presence of the surrounding ions. In non-metallic ionic crystals, in which the magnetic moments are definitely localized at the cations, the effect of the environment may be treated by crystal field theory. In metals, however, some of the electrons are delocalized, losing their identity with any specific ion core and belonging to an energy band which is common to the crystal as a whole. The magnetic moments may still be largely localized, but the moments of the ion cores in a crystal are very different from those of free atoms in a metal. This account will deal first with ionic solids, and sections on metals will follow.

Secondly the effects of short-range interactions become important in solids. These exchange interactions have so far simply been described by a hypothetical exchange field, but such a description makes no contribution towards their explanation. The interactions, if weak, merely modify the paramagnetic susceptibility but strong interactions, either between neighbouring ions (direct exchange) or via intervening anions (indirect exchange), may lead to parallel or antiparallel alignment of the moments, that is to ferromagnetic or antiferromagnetic ordering, provided the lattice is below a critical temperature. In view of the great range and complexity of ordered magnetic materials considerable selectivity is necessary, and attention will be concentrated on spinel ferrites and a few other ionic crystals.

4.1 CRYSTAL FIELDS

Crystal field effects apply to both paramagnetic and magnetically ordered materials. They may be introduced in a simple semi-graphical manner.

For an electron of a free ion, specified by the quantum numbers n and l, unless $l = 0$ there will be a number of states with equal energy. These degenerate states are represented by the different wavefunctions for different values of $m = l, l-1, ..., 0, ..., -(l-1), -l$ and so far the lifting of this degeneracy has been associated only with the effect of an applied magnetic field.

In the previous chapter the function $e^{im\phi}$ was shown to be an eigen-function of the operator for a component of the angular momentum, namely of $(\hbar/i)\partial/\partial\phi$, since it clearly gives the eigenvalues $m\hbar$. The angular parts of the wavefunctions, given in Table 3.1, contained $e^{im\phi}$ as the only function of ϕ and thus these were themselves eigenfunctions of $(\hbar/i)\partial/\partial\phi$, and represented states for which there were definite values of the components of the angular momentum. However, these are not the only permissible wavefunctions, and may themselves be unacceptable in certain situations.

The wavefunction as a whole must satisfy the Schrödinger equation

$$\hat{H}\psi = E\psi \tag{4.1}$$

which can be solved, after finding an appropriate expression for $\hat{H}$, so long as $\psi(r,\theta,\phi)$ is assumed to be separable into the product of functions of r, θ and ϕ, that is

$$\psi(r,\vartheta,\phi) = R(r)\Theta(\theta)\Phi(\phi). \tag{4.2}$$

This condition permits the separation of the Schrödinger equation into three ordinary differential equations each containing a single variable[1]. One of these equations is

$$\frac{d^2\Phi}{d\phi^2} + k^2\Phi = 0 \tag{4.3}$$

which has the solutions

$$\Phi = e^{ik\phi}$$

$$\Phi = e^{-ik\phi} . \tag{4.4}$$

If k is identified with m then one of these is an eigenfunction of the operator $\hat{m}_z$, or in other words the wavefunction as a whole is an eigen-function.

---

[1] This principle may be illustrated by taking a simpler example. Laplace's equation in two dimensions is

$$\frac{\partial^2 V}{\partial x^2} + \frac{\partial^2 V}{\partial y^2} = 0 .$$

Assume $V(x,y) = X(x)Y(y)$ and partially differentiate the product twice with respect to x and to y, remembering that $\partial Y/\partial x = \partial X/\partial y = 0$, which gives

$$Y\frac{\partial^2 X}{\partial x^2} + X\frac{\partial^2 Y}{\partial y^2} = 0 .$$

Divide by XY

$$\frac{\partial^2 X}{\partial x^2}\frac{1}{X} = -\frac{\partial^2 Y}{\partial y^2}\frac{1}{Y} .$$

If we assume that the two terms do not represent identical functions then each must be equal to a constant, which gives the two ordinary differential equations

$$\frac{d^2 X}{dx^2} = k^2 X \qquad \frac{d^2 Y}{dy^2} = -k^2 Y$$

with obviously simple solutions.

However, it is readily shown by substituting into the equation that any linear combination of these two solutions, of the form $\pm a_1 e^{im\phi} \pm a_2 e^{-im\phi}$ where a_1 and a_2 are constants, is also a solution. Particular examples are

$$\Phi = e^{im\phi} + e^{-im\phi} = 2\cos m\phi$$

$$\Phi = e^{im\phi} - e^{-im\phi} = 2\sin m\phi. \tag{4.5}$$

which are *not* eigenfunctions of $\hat{m}_z$, for example

$$\frac{\hbar}{i} \frac{\partial}{\partial \phi} (2\cos m\phi) = -\frac{2m\hbar}{i} \sin m\phi$$

For a 3d electron, $l = 2$; $m = 2, 1, 0, -1, -2$, and one obtains an alternative acceptable set of *real* wavefunctions as

$$\psi_{xy} = R(r)Y_{xy} = R(r)\sin^2\theta \sin 2\phi = \frac{R(r)}{r^2} xy$$

$$\psi_{x^2-y^2} = R(r)Y_{x^2-y^2} = R(r)\sin^2\theta \cos 2\phi = \frac{R(r)}{r^2}(x^2-y^2)$$

$$\psi_{yz} = R(r)Y_{yz} = R(r)\sin\theta \cos\theta \sin\phi = \frac{R(r)}{r^2} yz \tag{4.6}$$

$$\psi_{zx} = R(r)Y_{zx} = R(r)\sin\theta \cos\theta \cos\phi = \frac{R(r)}{r^2} zx$$

$$\psi_{z^2} = R(r)Y_{z^2} = R(r)(3\cos^2\theta - 1) = \frac{R(r)}{r^2}(2z^2-x^2-y^2).$$

These are sometimes designated d_{xy}, $d_{x^2-y^2}$, etc. and may be represented pictorially by the orbitals in Figure 4.1, which are obtained as the surfaces formed by the tips of radius vectors drawn in any direction (θ, ϕ) with lengths proportional to $Y(\theta, \phi)^2$ in that direction. Thus they represent the angular distribution of the electron density (if the orbitals are in fact occupied), the complete electron density distributions being given by including the effect of the radial functions, $R(r)^2$ and the scale being given by the normalization constants, N. The orbitals fall into two different groups on the grounds of their symmetry, the group d_{xy}, d_{yz} and d_{zx} having a symmetry designated either t_{2g} or d_ϵ and $d_{x^2-y^2}$ and d_z^2 having the symmetry d_γ or e_g.

Now suppose that the ion is placed in a crystal lattice. One of the most frequently occurring types of site, in oxides at least, is an octahedral site or one with octahedral coordination, that is surrounded by six anions (such as O^{2-}) symmetrically disposed as in a cubic crystal structure. (Drawing plane surfaces through the anions forms a regular octahedron: a tetrahedral site is one in the centre of a tetrahedron formed by four anions.) The simplest way to introduce the effect of the anions is to represent them as point negative charges on the rectangular

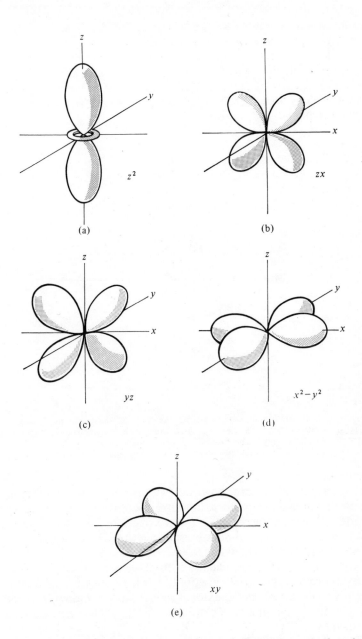

Figure 4.1. Pictorial representation of the d orbitals ($l = 2$) showing the angular dependence of the possible electron distributions, as obtained from Equation (4.6). The radial distribution is not represented.

axes as shown in Figure 4.2 (although in fact this is only a permissible formality since the real effects are stronger than those calculated on purely electrostatic grounds). The fields from the charges will influence the energies of the different orbitals in different ways. For a d^1 ion the single electron is concentrated near to the negative charges for the d_γ (e_g) orbitals, and these will be raised in energy as compared with the three equivalent t_{2g} (d_ϵ) orbitals.

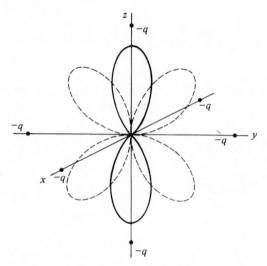

Figure 4.2. A d^1 cation in an octahedral site, the six neighbouring anions being represented by point negative charges on the rectangular axes. The full line represents a d_{z^2} orbital, which is raised in energy. The broken line represents a d_{yz} orbital, the energy associated with which is lowered by the interaction as compared with the d_{z^2} orbital.

The first point to note, however, is that when the crystal fields with this particular symmetry are taken into account it can be demonstrated (Griffiths [1]) that the set of $3d^1$ wavefunctions given in Equation (4.6) are in fact acceptable and do represent relatively low energies. The type of theory applied is known as *perturbation theory*, the perturbation in this case arising from the inclusion of the electrostatic potential in the energy Hamiltonian to give an equation which cannot be solved exactly. It may also be noted that the direction of spin is not directly affected by electrostatic fields, and so we are justified in neglecting spin entirely at this stage.

Thus, the crystal fields may be considered to have three interrelated effects. They define a particular set of wavefunctions; they lift the degeneracy of the energy levels, and they lead to the 'quenching' of the orbital angular momentum, as described in section 4.1.2.

4.1.1 Crystal field splitting

The changes in the energy levels of an electron by a perturbation which adds a term $\hat{H}_p$ to the Hamiltonian, giving $\hat{H}_0 + \hat{H}_p$ where $\hat{H}_0$ applies to the isolated ion, are found by carrying out the integration

$$\Delta E = \iiint \psi^* \hat{H}_p \psi \, d\tau \qquad (4.7)$$

over all space. When this is done it is found, for the present case, that the splitting can be represented to a good approximation in terms of a coefficient D which appears in the classical expression for the electrostatic potential at an octahedral site:

$$V = \text{constant} + D(\text{4th order terms}). \qquad (4.8)$$

The calculation gives the splitting in terms of this crystal field constant and a parameter q, which is the mean value of r^4 in $R(r)$, as

$$\Delta E = +6Dq \quad \text{for } d_\gamma$$

and

$$\Delta E = -4Dq \quad \text{for } d_\epsilon, \qquad (4.9)$$

but it must be remembered that this is only an approximate treatment because of the classical representation of the effects of the anions.

Inspection of Figure 4.1 shows that it is not surprising that the three orbitals d_{xy}, d_{yz}, and d_{zx} should be affected in an identical manner. Also, the two d_γ orbitals should be affected in an opposite sense, and the fact that they are equally affected follows from the possibility of equating ψ_{z^2} to a linear combination of $\psi_{y^2-z^2}$ and $\psi_{z^2-x^2}$, which have the same symmetry as $\psi_{x^2-y^2}$.

The splitting can be represented diagrammatically by the energy level diagram of Figure 4.3a. The upper level is still doubly degenerate and the lower level triply degenerate. For a $3d^9$ ion the situation is reversed, since a full shell gives a spherical charge cloud and the loss of one electron gives rise effectively to positively charged orbitals for the 'electron hole'. Thus the diagram for an ion such as Cu^{2+} is as shown in Figure 4.3b.

For a half-filled shell, $3d^5$, $L = 0$ (see Figure 3.4) and there can be no splitting for this S state (Figure 4.3c). Neglecting the spherically symmetrical distribution for the first 5 electrons a $3d^6$ ion is equivalent to $3d^1$, and similarly the effect on d^4 is equivalent to that on d^9.

The remaining degeneracy may be largely lifted if the environment is not perfectly octahedral. The simplest case is represented by a uniaxial component of the crystal field, superimposed on the cubic field, as may arise from Jahn–Teller distortion. Jahn and Teller pointed out that a distortion of the lattice could cause the lower orbital multiplet to be split in such a way that the lowest singlet then has a lower energy than the multiplet; thus, the net energy of the system can be lowered by a spontaneous distortion of the lattice. The effect of such a distortion for Cu^{2+} is shown in Figure 4.4.

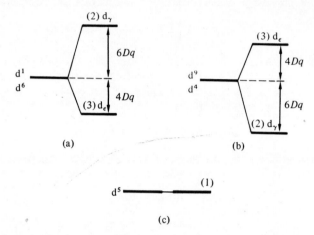

Figure 4.3. (a) A d^1 ion has $L = l = 2$, $M_l = 2, 1, 0, -1, -2$; that is it has five-fold degeneracy in isolation. In an octahedral crystal field this degeneracy is partially lifted, since the two d_γ orbitals are raised in energy while the three d_ϵ orbitals are lowered in energy as indicated. The absolute value of the energy is arbitrary, but the splittings are as shown.

(b) The crystal field splittings for a d^9 or d^4 ion, being the reverse of those in 4.3.(a).

(c) A d^5 ion has $L = 0$, $M_L = 0$, that is it has no degeneracy and a spherically symmetrical electron distribution; thus there can be no crystal field splitting.

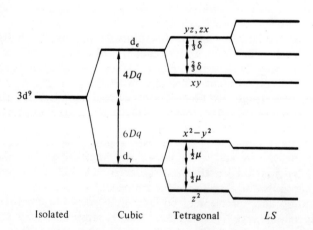

Figure 4.4. The cubic (octahedral) crystal field splitting is as shown in Figure 4.3.(b). A tetragonal distortion of the lattice gives a crystal field of lower symmetry, and the degeneracy of the lower level is lifted. The remaining degeneracy is lifted by the spin-orbit coupling, LS.

The figure also indicates the removal of the remaining degeneracy by the effect of the spin-orbit coupling. As noted in section 3.10 this coupling governs the relative orientations of **L** and **S**, and is represented by an additional term in the Hamiltonian, $\hat{H}_{so} = \lambda \mathbf{L} \cdot \mathbf{S}$. A question which may have occurred to the reader by now is, why has the spin-orbit coupling not been introduced at the beginning of the discussion, or in other words why did we not consider the effect of the crystal fields on the total angular momentum $\mathbf{J} = \mathbf{L} + \mathbf{S}$? The answer is provided by consideration of the magnitudes of the different energy terms involved, as represented by the contributions to the Hamiltonian in Table 4.1.

Table 4.1 (all values expressed as 10^4 cm^{-1}).

	$\hat{H}_e$ (electrostatic)	$\hat{H}_c$ (cubic field)	$\hat{H}_t$ (tetragonal)	$\hat{H}_{so}$ (spin-orbit)	$\hat{H}_H$ (mag. field)
3d	2	1–2	0·01	0·02	1×10^{-6}
4d	2	2–4	0·01	0·1	1×10^{-6}
5d	2	2–4	0·01	0·1	1×10^{-6}
4f	2	< 0·1		0·1	1×10^{-6}

The following conclusions may be drawn:

1. For the 3d transition series the crystal field effects overcome the spin-orbit coupling, thus showing that the lowest energy level (which will be almost wholly the occupied level if it is distant from the next highest level by more than $kT = 200$ cm^{-1} at room temperature) is governed by L and S rather than J; hence J is not a good quantum number.

2. For the 4f ions the spin-orbit coupling predominates over the crystal field energy, and J is a good quantum number. The crystal fields have a secondary effect in this case, just as the spin-orbit coupling has a secondary effect in our first conclusion. However, it is immediately apparent why the 4f ions in crystals behave approximately as free ions, and why, in Table 3.2, $p_e \doteqdot g[J(J+1)]^{1/2}$. Qualitatively the distinction between 3d and 4f ions arises from the observation that the electrons, which give rise to the magnetic moments in the latter case, are largely screened from the crystal fields by the external 5s and 5p electrons.

3. For the 4f and most of the 3d ions the electrostatic or coulomb energy, $\hat{H}_e$, predominates and the coupling of the individual electron spins to give ionic spin S, and the individual l's to give L, is not affected by the crystal fields.

4. For some of the 3d ions, particularly in the strong crystal fields or ligand fields which arise in complexes such as ferricyanides, and for some of the other transition series ions (4d and 5d), $\hat{H}_e$ does not predominate and thus certain environments can modify the application of Hund's rules described in section 3.9. The individual electrons may then be assigned the lowest orbital momenta, since the spins are not directly

affected by the crystal fields. To fill the 3d shell, the first 3 electrons will occupy the d_ϵ level and, since this is triply degenerate, they may have parallel spins in the usual way; however, instead of the electrostatic coupling enforcing a parallel spin for the fourth electron, so that it must then by the Pauli principle enter the d_γ level, the large spacing of the levels causes the fourth electron also to enter the d_ϵ level and so the spin must be antiparallel. The total spin is then less than would otherwise be expected, that is a *low spin state* is induced. This can be seen to apply to the ions from d^4 to d^7.

4.1.2 Quenching of the orbital angular momentum

In view of the foregoing the orbital and spin angular momenta of 3d ions in strong crystal fields may initially be treated separately. (The secondary spin-orbit coupling will then constitute a correction to this approach.)

It has been noted that the wavefunctions are not eigenfunctions of $\hat{m}_z$. It is a basic postulate of quantum theory that the mean value of repeated observations, represented by the operator $\hat{A}$, on a state with wavefunction ψ is

$$\bar{a} = \frac{\int \psi^* \hat{A} \psi \, d\tau}{\int \psi^* \psi \, d\tau}, \tag{4.10}$$

the denominator often being omitted on the assumption that the wavefunction is normalized. This is clearly relevant if ψ is an eigenfunction of A, because then $\hat{A}\psi = a_n \psi$ and

$$\bar{a} = a_n \int \psi^* \psi \, d\tau = a_n, \tag{4.11}$$

that is the observation always gives the eigenvalue a_n.

For the case in question only $\Phi(\phi)$ need be considered, and this will be taken as $\sin m\phi$ so that the mean value of the component of the angular momentum is

$$\begin{aligned}
\bar{l}_z &= \int_{\phi=0}^{\phi=2\pi} (\sin m\phi) \frac{\hbar}{i} \frac{\partial}{\partial \phi}(\sin m\phi) \, d\phi \\
&= \int_0^{2\pi} (\sin m\phi) \frac{\hbar}{i} m(\cos m\phi) \, d\phi \\
&= \frac{\hbar m}{im} \int_0^{2\pi} (\sin m\phi) \, d(\sin m\phi) \\
&= \frac{\hbar}{2i} [\sin^2 m\phi]_0^{2\pi} = 0
\end{aligned} \tag{4.12}$$

Thus, although the orbital angular momentum exists it cannot have any finite component in any given direction nor make any contribution to the atomic magnetic moment. The extremely important and simple conclusion reached is that for the 3d ions in crystals, or in the ligand fields arising in ionic molecules, *the magnetic moment is contributed by the spin only*. Hence, in a paramagnetic the atomic moment should be given by

$$\mu = g\mu_B [S(S+1)]^{\frac{1}{2}} \doteqdot 2\mu_B [S(S+1)]^{\frac{1}{2}}.$$

That this is approximately so is seen by reference to Table 3.2. The deviations from this rule are discussed in section 4.4 in relation to anisotropic properties.

The components of the magnetic moment are given even more simply as $2\mu_B S$ or $n\mu_B$, where n is the number of unpaired spins. The susceptibility may be calculated very briefly for this case. An applied field may generally be assumed to cause a splitting of the (spin degenerate) ground state energy level, given by $\Delta E = 2\mu \cdot H = 2g\mu_B SH$, which is of the order of 1 cm^{-1} and is thus much less than the separation of the first excited state and also very much less than kT. If the population of the lower spin state ($\mu \parallel H$) is N_1, and that of the higher spin state is N_2 (where $N_1 + N_2 = N$), then

$$\frac{N_1}{N} = \frac{e^x}{e^x + e^{-x}},$$

$$\frac{N_2}{N} = \frac{e^{-x}}{e^x + e^{-x}}$$

where $x = \mu H/kT \ll 1$, so that $e^x \doteqdot 1 + x$ and

$$N_1 = \tfrac{1}{2}N(1+x),$$

$$N_2 = \tfrac{1}{2}N(1-x).$$

The magnetization per unit volume is

$$I = (N_1 - N_2) \times \text{(atomic moment)}$$

$$= (N_1 - N_2)g\mu_B S$$

$$= Ng\mu_B S \frac{g\mu_B SH}{kT} \tag{4.13}$$

$$\chi = \frac{I}{H} = \frac{Ng^2\mu_B^2 S^2}{kT} = \frac{C}{T}. \tag{4.14}$$

The ordering in ferromagnetic materials etc. has been represented by an exchange field, and so the saturation magnetization is calculated from the vector sum of the M_S values. For simple ferrimagnetic crystals the calculation requires a knowledge of the population of the two anti-parallel sublattices.

4.2 EXCHANGE MECHANISMS

Exchange interactions or exchange energies, which have been described as the fundamental basis of ferromagnetism and antiferromagnetism, are directly associated with the proximity of the ions in crystals or in molecules. Direct exchange requires overlap of the wavefunctions, which have a limited extent, indicated by f(r). Indirect exchange, as discussed later, requires overlap between the wavefunctions of the two cations considered and an intervening anion.

The degree of overlap of any two atomic wavefunctions can be represented by the overlap integral

$$\alpha = \int \psi_a^*(\mathbf{r})\psi_b(\mathbf{r})\,d\tau. \tag{4.15}$$

It is clear that this expression has the appropriate properties of being zero when the origins of the wavefunctions (that is the nuclei) are so far apart that ψ_a is zero wherever ψ_b is finite and vice versa; also $\alpha = 1$ if the origins coincide so that $\psi_b \equiv \psi_a$ and Equation (4.15) just represents the normalization condition.

Whenever there is overlap as so indicated, which is, in effect, when the spacing of two nuclei is comparable with the atomic diameter as estimated from crystal lattice spacings, kinetic phenomena in gases etc., the energy of the system as a whole is found to contain terms that are additional to the simple long-range magnetostatic or electrostatic effects. (These latter are in fact too weak to be of great account in most cases.) If, for example, we accept that it is possible to calculate the energies of interaction of two hydrogen atoms coming together to form a molecule, it is to be expected that they will be a function of the internuclear spacing. In fact the results of the quantum mechanical calculations are as shown in Figure 4.5.

The upper curve represents the case when the spins of the two electrons are parallel, and the lower curve is for antiparallel spins. Thus, the total spin in the first case is $S = \frac{1}{2}+\frac{1}{2} = 1$; $2S+1 = 3$ and the molecule is said to be in a triplet state, while the antiparallel spins correspond to a singlet state, $S = 0$, $2S+1 = 1$. There is a minimum in the energy of interaction for the singlet state only, so a stable molecule will only form with antiparallel spins (the bonding state) representing a sort of rudimentary antiferromagnetism. Needless to say, molecular hydrogen is diamagnetic and the formation of molecules most generally involves such spin pairing, although there are exceptions as indicated by the paramagnetism of oxygen, O_2.

It is a large step from the hydrogen molecule to a crystal and, in particular, to a ferromagnetic crystal. However, it is instructive to sketch the manner in which the calculations leading to the results given in the figure may be made, simply because these calculations can be carried out with fair confidence, whereas more and more approximations are involved the more complex the system.

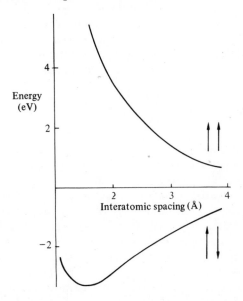

Figure 4.5. The approximate form of the total energy of a hydrogen molecule as a function of interatomic spacing, calculated for parallel or antiparallel spins of the two electrons as indicated. An energy minimum only occurs for antiparallel spins.

It is assumed, in Heitler–London or molecular orbital methods, that the wavefunction for the two electrons $\psi(\mathbf{r}_1,\mathbf{r}_2)$ can be constructed as a linear combination of the two wavefunctions for the single electrons in the separate atoms. For brevity represent the coordinates of one electron $(x,y,z$ or $r,\theta,\phi)$ by $\mathbf{r}_1$, and the position of the second electron by $\mathbf{r}_2$. If electron 1 is in a state 'a' its wavefunction can be written $\psi_a(\mathbf{r}_1)$, and the atomic wavefunction of the other electron, in state 'b' may be written $\psi_b(\mathbf{r}_2)$. Consider the possible combinations

$$\psi(\mathbf{r}_1,\mathbf{r}_2) = N\{\psi_a(\mathbf{r}_1)\psi_b(\mathbf{r}_2)\} \qquad (4.16a)$$

and

$$\psi(\mathbf{r}_1,\mathbf{r}_2) = N_\pm\{\psi_a(\mathbf{r}_1)\psi_b(\mathbf{r}_2) \pm \psi_b(\mathbf{r}_1)\psi_a(\mathbf{r}_2)\}. \qquad (4.16b)$$

For the first combination, assuming that a and b represent different states, the interchange of $\mathbf{r}_1$ and $\mathbf{r}_2$ changes the value of $\psi(\mathbf{r}_1,\mathbf{r}_2)^2$, that is it alters the electron distribution at all points. However, the designation of the two electrons by separate subscripts is a matter of convenience only: we know that there are two electrons in the system but there is no conceivable experiment that could be performed which would identify each of them separately. Thus, the interchanging of $\mathbf{r}_1$ and $\mathbf{r}_2$ in our description of the system must not affect that description in any way which would indicate that different observations could be made. This condition is met by the two alternatives contained in Equation (4.16b). With the + sign $\psi(r_1, r_2)$ remains exactly the same after the interchange and in such a case ψ is called *symmetric*; but with the − sign ψ^2 also remains unchanged because ψ simply changes sign (the two products replacing each other) and ψ is then called *antisymmetric*. The antisymmetric functions can be written in the form of a determinant

$$\psi(r_1, r_2) = N_- \begin{vmatrix} \psi_a(r_1) & \psi_b(r_1) \\ \psi_a(r_2) & \psi_b(r_2) \end{vmatrix} \tag{4.17}$$

Note that one of the basic properties of determinants is that their value is zero if any two rows or columns are identical. Now let us suppose that, instead of dealing only with the spatial coordinates, or orbital part of the wavefunction, we had used wavefunctions containing information on the spins of the electrons. This can be done by the introduction of spin coordinates which, in the presence of a magnetic field defining a particular direction, can have only two values, $\pm\hbar/2$, since the electron spin can only be parallel or antiparallel to the field. If the wavefunctions for the spin are designated χ_α and χ_β for the two possible spin states of the one electron, then acceptable expressions will be in the form of matrices

$$\chi_\alpha = \begin{pmatrix} 1 \\ 0 \end{pmatrix}, \qquad \chi_\beta = \begin{pmatrix} 0 \\ 1 \end{pmatrix}; \tag{4.18}$$

if the operator for the spin angular momentum is taken to be the Pauli spin operator

$$\hat{s}_z = \frac{\hbar}{2} \begin{pmatrix} 1 & 0 \\ 0 & -1 \end{pmatrix}, \tag{4.19}$$

for then we have

$$\hat{s}_z \chi_\alpha = \frac{\hbar}{2} \begin{pmatrix} 1 & 0 \\ 0 & -1 \end{pmatrix} \begin{pmatrix} 1 \\ 0 \end{pmatrix} = \frac{\hbar}{2} \begin{pmatrix} 1 \\ 0 \end{pmatrix} = \frac{\hbar}{2} \chi_\alpha. \tag{4.20}$$

By an obvious analogy with the differential operators used in the previous chapters one can see that the correct eigenvalues are produced by this scheme. The complete wavefunctions, including both orbital and spin contributions, can now be written as the product

$$\Psi = \psi(r)\chi \tag{4.21}$$

(if the orbital angular momentum is quenched, so that the orbit and spin can be considered separately) since acting on this product with the spin operator will still just give the spin angular momentum components. Now, if we return to Equation (4.17) and assume that each of the functions ψ is multiplied by the same spin function χ, it can be seen why in practice the complete wavefunctions must always be antisymmetric, because then the opening note to this paragraph becomes consistent with the Pauli exclusion principle according to which the probability of finding two electrons in the same state is zero. However the functions $\psi(\mathbf{r}_1, \mathbf{r}_2)$ alone can be either symmetric or antisymmetric.

We now return to the consideration of the orbital functions only, postponing further consideration of the spin. The important calculation to be made is that of the two energies connected with symmetrical and antisymmetrical states, which will be given now by a double integral (cf. section 4.1.2)

$$E = \iint \psi^*(\mathbf{r}_1, \mathbf{r}_2)\hat{H}\psi(\mathbf{r}_1, \mathbf{r}_2)\,d\tau_1 d\tau_2. \tag{4.22}$$

The Hamiltonian will contain the Hamiltonians for each electron $\hat{H}_1$ and $\hat{H}_2$, and also a term to account for the interaction, that is

$$\hat{H} = \hat{H}_1 + \hat{H}_2 + \hat{H}_{12}.$$

$\hat{H}_1$ contains only the coordinate $\mathbf{r}_1$ and $\hat{H}_2$ only $\mathbf{r}_2$, and when the interaction is neglected each gives rise to the appropriate wavefunctions according to the Schrödinger equations

$$\hat{H}_1 \psi_a(\mathbf{r}_1) = E_a \psi_a(\mathbf{r}_1)$$

$$\hat{H}_2 \psi_b(\mathbf{r}_2) = E_b \psi_b(\mathbf{r}_2). \tag{4.23}$$

The solution of Equation (4.22), after deriving a suitable Hamiltonian (as by Heitler and London) is rather lengthy and the results must be quoted as

$$E = E_a + E_b + 2N_\pm^2(Q \pm J), \tag{4.24}$$

in other words, they depend on whether the wavefunction is symmetric or antisymmetric. E_a and E_b are as defined by Equation (4.23). Q is called the *Coulomb energy* and J the *exchange integral*:

$$Q = \iint \psi_a^*(\mathbf{r}_1)\psi_b^*(\mathbf{r}_2)\hat{H}_{12}\psi_a(\mathbf{r}_1)\psi_b(\mathbf{r}_2)\,d\tau_1 d\tau_2, \tag{4.25}$$

$$J = \iint \psi_a^*(\mathbf{r}_1)\psi_b^*(\mathbf{r}_2)\hat{H}_{12}\psi_b(\mathbf{r}_1)\psi_a(\mathbf{r}_2)\,d\tau_1 d\tau_2. \tag{4.26}$$

The coefficient $N_\pm$, the normalizing constant in Equation (4.16b), introduces the overlap [Equation (4.15)] since

$$N_\pm^2 = \frac{1}{2(1 \pm \alpha^2)}. \tag{4.27}$$

The reader should note how J arises from the necessity of choosing symmetrical or antisymmetrical wavefunctions. If the first of the two choices in Equation (4.16) had been made, the value of Q would have been the same, but J would not have arisen at all. The choice was made by considering the effect of interchanging the electrons, and thus the real origin of J lies in the same argument—hence the term exchange energy. [Note also that the exchange energy splitting is derived from the exchange integral by $E_J = 2J(N_+^2 - N_-^2)$ according to Equation (4.24)].

We have yet to explain the way in which the alignment of the spins follows from these orbital interactions. Further symmetry arguments are used. It has been seen that spin eigenfunctions are $\chi_\alpha(\sigma)$ and $\chi_\beta(\sigma)$, where σ represents the spin coordinate which can have the values ± 1 only. An arbitrary spin state for two electrons can be represented by simple linear combinations of these eigenfunctions, in a manner closely comparable to that of the case of the spatial functions. Once again the combinations must have the property of being either symmetric or antisymmetric, and inspection shows that the four two-electron spin functions satisfy this condition:

$$\chi_A(\sigma_1, \sigma_2) = \chi_\alpha(\sigma_1)\chi_\beta(\sigma_2) - \chi_\beta(\sigma_1)\chi_\alpha(\sigma_2), \tag{4.28}$$

$$\chi_s^I(\sigma_1, \sigma_2) = \chi_\alpha(\sigma_1)\chi_\beta(\sigma_2) + \chi_\beta(\sigma_1)\chi_\alpha(\sigma_2), \tag{4.29}$$

$$\chi_s^{II}(\sigma_1, \sigma_2) = \chi_\alpha(\sigma_1)\chi_\alpha(\sigma_2), \tag{4.30}$$

$$\chi_s^{III}(\sigma_1, \sigma_2) = \chi_\beta(\sigma_1)\chi_\beta(\sigma_2). \tag{4.31}$$

These three symmetric and one antisymmetric functions constitute the only four possibilities. The spin operator for two electrons can be written as $\hat{s}_{z,1} + \hat{s}_{z,2}$ with the implication that $\hat{s}_{z,1}$ acts only on the spin coordinate σ_1 and $\hat{s}_{z,2}$ on σ_2. If we apply this condition to χ_s^{II} for example:

$$(\hat{s}_{z,1} + \hat{s}_{z,2})\chi_s^{II} = (\hat{s}_{z,1} + \hat{s}_{z,2})\chi_\alpha(\sigma_1)\chi_\alpha(\sigma_2)$$

$$= (\hbar/2)\chi_\alpha(\sigma_1)\chi_\alpha(\sigma_2) + (\hbar/2)\chi_\alpha(\sigma_1)\chi_\alpha(\sigma_2)$$

$$= +1\hbar\chi_s^{II}, \tag{4.32}$$

that is $M_s = 1$ and the spins are parallel for this state. It can similarly be shown (using the operator for the magnitude as well as the components of the spin angular momentum) that the values of M_s and S, where $\mathbf{S} = \hbar[S(S+1)]^{1/2}$, are as follows:

χ_A	$M_s = 0$	$S = 0$
χ_s^I	$M_s = 0$	$S = 1$
χ_s^{II}	$M_s = 1$	$S = 1$
χ_s^{III}	$M_s = -1$	$S = 1.$

χ_A represents the singlet state (antiparallel spins) and χ_s the triplet state (parallel spins). Now since the total wavefunction must be antisymmetric with respect to the interchange of $(\mathbf{r}_1, \sigma_1)$ and $(\mathbf{r}_2, \sigma_2)$, of the four possible combinations

$$\Psi = \psi_A(\mathbf{r}_1, \mathbf{r}_2) \chi_A(\sigma_1, \sigma_2), \qquad (4.33)$$

$$\Psi = \psi_s(\mathbf{r}_1, \mathbf{r}_2) \chi_s(\sigma_1, \sigma_2), \qquad (4.34)$$

$$\Psi = \psi_A(\mathbf{r}_1, \mathbf{r}_2) \chi_s(\sigma_1, \sigma_2), \qquad (4.35)$$

$$\Psi = \psi_s(\mathbf{r}_1, \mathbf{r}_2) \chi_A(\sigma_1, \sigma_2), \qquad (4.36)$$

inspection shows that only the latter two qualify. In other words, if ψ is symmetric, χ must be antisymmetric and vice versa. Thus, since symmetric and antisymmetric χ's correspond to parallel and antiparallel spins, the same energy splitting which separates the symmetric and antisymmetric orbital levels applies also to the splitting between parallel and antiparallel spins. To this extent the exchange energy, as already derived, can be taken as an interaction between the spins. For parallel spins (Equation, 4.35) the exchange energy is $2N_+^2(-J)$, and for antiparallel spins it is $2N_+^2(+J)$, and so a positive value for J indicates that the parallel spin configuration has the lower energy, and vice versa.

For the above reasons an exchange Hamiltonian may be used to represent the effective spin interaction, namely

$$\hat{H}_e = -2J_{ij}\mathbf{s}_i \cdot \mathbf{s}_j$$

or

$$\hat{H}_e = -2J_{ij}\mathbf{S}_i \cdot \mathbf{S}_j, \qquad (4.37)$$

since $\mathbf{S} = \sum \mathbf{s}$ when there is more than one electron per ion. If the exchange is assumed to be isotropic, all the interactions are the same and

$$\hat{H}_e = -2J_{ej}\sum \mathbf{S}_i \cdot \mathbf{S}_j \qquad (4.38)$$

can be taken to represent the exchange interaction of the spin of one ion $\mathbf{S}_i$, with all its neighbours.

4.2.1 Ionic exchange interactions

The great majority of magnetically ordered ionic crystals are antiferromagnetic (ferrimagnetism being considered as a special case of this). However, a limited number of ionic compounds are known to be ferromagnetic, generally with low Curie points (e.g. EuO, $CrBr_3$, CrO_2).

The typical situation may be represented by a basic model of two cations with an intervening anion, for example $Mn^{2+} O^{2-} Mn^{2+}$. Apart from the presence of the intervening anion the direct spacing of the cations will be such as to preclude any very great overlap, although there is a possibility of some direct exchange in a non-linear sense (for example the cation orbitals may overlap along a line which does not join their centres directly but contains a $90°$ angle). Thus, the primary coupling is indirect and is effected via the intermediary O^{2-} ion. The first demonstration of this appears to have been due to Anderson [2], following Kramer's principle that the oxygen will be partially in an excited state, that is a ground state anion p electron may be excited into a d state on a cation. Suppose that it enters the next available level in the 3d shell: if this is half full or more the excited spin must be antiparallel to the net 3d spin, whereas if the shell is less than half full all the d electron spins will become parallel. The spin of the remaining p electron will still be coupled antiparallel to the excited spin, and thus either parallel or antiparallel to the d spins of cation 2 (Figure 4.6). Since the oxygen, which has lost an electron, has a net spin it can be exchange coupled by means of direct overlap with the spin of cation 1, and this coupling is expected to be antiferromagnetic in nature. The overall coupling between the cations depends on a combination of direct exchange, excitation, and intra-atomic (Hund's rule) coupling, and is known as *super exchange*. The effect will be to give ferromagnetism or antiferromagnetism according to whether the cation d shells are less than half full or at least half full, and it is clear that this does occur in some cases. It corresponds to the antiferromagnetism of MnO, NiO, CoO, FeO etc., while a few compounds such as CrO_2 and CrTe are indeed ferromagnetic (Cr^{2+} is $3d^4$).

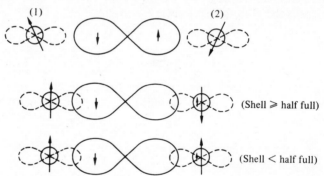

Figure 4.6. Super exchange, or Kramers–Anderson mechanism, showing the initial state and the final state for cation shells at least half full or less than half full.

Zener's double exchange theory [3] involves the excitation of a d electron from the cation with the highest number of such electrons (lower valency) into an overlapping anion orbital, with the transfer of one anion p electron to the other cation. As shown in Figure 4.7 the cations may effectively interchange positions and become ferromagnetically coupled, but it follows that this cannot occur when the cations are ordered in the crystal lattice according to their valency.

The formation of hybrid orbitals by combinations of d, s, and p orbitals in such a way as to give partial covalent bonding has been considered by Goodenough. The p electrons can spend part of their time in these vacant hybrid orbitals, without change of spin, and give a resultant coupling between the cations as illustrated by Figure 4.8. This is known as *semi-covalent exchange*.

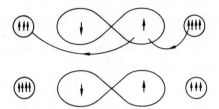

Figure 4.7. Double exchange in which an electron is effectively transferred from one cation to another.

(a)   (b)

Figure 4.8. (a) The overlap of atomic orbitals of next-nearest neighbour cations with an intervening anion forming hybrid orbitals $\psi = d + \delta p$ where δ is related to the overlap integral—Equation (4.15). The 90° interaction between nearest neighbours involves overlap with different p orbitals and gives a positive exchange interaction. Direct overlap between cation orbitals is sometimes possible, as for B site cations in spinels, giving negative direct exchange (b).

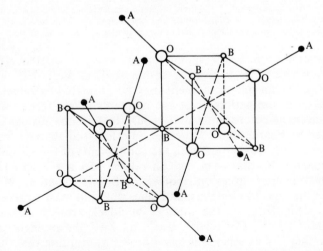

Figure 4.9. A part of the spinel crystal lattice, which applies to the ferrimagnetic ferrites. The oxygen ions (O) form a cubic lattice and the cations can occupy either octahedral or B sites (○), or tetrahedral A sites (●): the octahedral symmetry of the central cation is particularly clear.

The reader should make further reference to the original papers by Goodenough [4] and Anderson [2], and also by Kanamori [5] and Blasse [6] for example, to see how these principles apply to oxides of special interest, particularly the spinel ferrites and rare-earth garnets. The spinel structure consists of a face-centred cubic lattice of oxygen containing two interpenetrating lattices of either octahedral, B, or tetrahedral, A, sites, as illustrated by Figure 4.9. Eight A sites and sixteen B sites are occupied per unit cell. In a normal spinel divalent ions occupy the A sites and trivalent ions the B sites, while in an inverse spinel eight trivalent ions occupy the A sites and the B sites contain eight each of trivalent and divalent ions. The inverse arrangement is the most common. When all the super exchange and direct exchange interactions are taken into account, including the 90° exchange (or the interaction between orbitals at 90° to each other rather than along a line including an anion), it is found that the cations in A and B sites are coupled in an antiferromagnetic manner. Consequently, all the A site ions are uniformly magnetized in the opposite direction to all the B site ions.

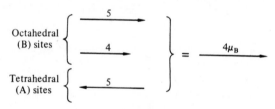

Figure 4.10. The arrangement of the ionic magnetic moments in a ferrite such as magnetite with inverse population: for each formula unit one each of the ferric ions is in each sublattice, so that the net moment is that of the divalent ion, $4\mu_B$ in the present case.

The general formula for spinel ferrites may be written MFe_2O_4 or $MOFe_2O_3$ where M is a divalent cation such as Fe^{2+}, Ni^{2+} etc. The spontaneous ferrimagnetic moment may be estimated extremely simply for the inverse occupation, since the moments of the ferric ions cancel and the moment per formula unit is just that of the divalent ions, as represented by Figure 4.10. Moreover, in calculating the vector magnetization one is only concerned with the components of the spin moments, not the magnitude, that is $m_s = \pm\frac{1}{2}\hbar$ rather than $s = \hbar[s(s+1)]^{1/2}$, and thus each 'unpaired' spin contributes one Bohr magneton. Mn^{2+} gives $5\mu_B$. Fe^{2+} has six 3d electrons, and since one spin must be reversed this gives $4\mu_B$. Similarly, as the number of 3d electrons increases to ten through Mn^{2+}, Fe^{2+}, Co^{2+}, Ni^{2+}, Cu^{2+} and Zn^{2+} the moment decreases from 5 to $0\mu_B$. This scheme is based on the assumption that the orbital contributions are substantially quenched, and the relatively good agreement with experimental values shows that this is so. Some properties of single ferrites are given in Table 4.2 (largely after Smit and Wijn [7]).

Table 4.2.

	n_B (calc.)	n_B	I_s	I_o	T_C °C	$K_1 \times 10^3$	$\lambda_s \times 10^6$	ρ (ohm cm)
$MnFe_2O_4$[a]	5	4·5	400	560	290–330	−40·7	−5	10^4
$FeFe_2O_4$	4	4·1	480	510	585	−130	+40	10^3
$CoFe_2O_4$	3	3·9	425	495	520	+2000	−110	10^7
$NiFe_2O_4$	2	2·3	270	300	570–600	−69	−17	10^4
$CuFe_2O_4$	1	1·3	130	160	455	−63	−10	10^5
$ZnFe_2O_4$	0	0						
$MgFe_2O_4$	0	1·2–2·2[b]	110	140	440	−25	−6	10^6
$Li_{0.5}Fe_{2.5}O_4$[c]	2·5	2·6	310[d]	330	590–680	−83[d]	−8	10^2

[a] Manganese ferrite is predominantly a normal spinel, but Mn^{2+} has the same magnetic moment as Fe^{3+}.
[b] The ionic distribution depends on the heat treatment and rate of cooling.
[c] Lithium is monovalent, and must be compensated by a certain proportion of Fe^{3+} ions.
[d] Varies with the degree of ordering which occurs in this ferrite.

It is particularly interesting to note that zinc ions, although having no moment themselves, increase the magnetization by decreasing the 'negative' or smaller sublattice magnetization. The commercial ferrites which are most common for use up to at least microwave frequencies are manganese–zinc and nickel–zinc compositions, which are referred to again in Chapters 5 and 6.

Some discrepancies in the table arise because not all the ferrites are completely inverse. In any case, the properties often depend on the method of preparation, slight departures from stoichiometry, etc., and the table should only be taken as an approximate guide: the ranges of values reported for a few of the figures stress this point.

4.3 METALS

The description of the magnetic properties of ionic solids is reasonably straightforward, due to the possibility of relating them fairly directly to the properties of the isolated ions. While the effects of the crystalline environment are very considerable there is no change in the basic structure of the ions; the number of electrons per ion remains unaltered and, apart from the promotion of electrons from cations in the super exchange mechanisms, the electrons remain strictly localized on their parent ions. This is why the calculation of ionic magnetic moments is so remarkably simple if complete quenching of the orbital momentum is assumed—one simply sums the unpaired spins in the d shell, for example, to give a strictly integral number of Bohr magnetons per ion for the component of the magnetic moment. (Of course there are some complications, such as the existence of high-spin and low-spin states.)

The treatment of metals may still start with that of the isolated ions, the 'tight binding approximation', but it soon becomes apparent that the ions in a metal crystal maintain these properties only to a very limited extent. The necessity for a special approach is stressed by two particular features: the existence of Pauli paramagnetism in some metals and the occurrence of non-integral ionic magnetic moments (in terms of numbers of Bohr magnetons) in the ferromagnetic metals, as indicated by Table 4.3. The neutral nickel atom, for example, is $3d^9$ and Ni^{2+} is $3d^8$, suggesting moments of one and two Bohr magnetons with completely quenched angular momentum. In fact, the spontaneous magnetization for nickel corresponds to about $0 \cdot 6\mu_B$ per ion, and it is confirmed by neutron diffraction that the 3d spin contributes $0 \cdot 656\mu_B$, the 3d orbit $0 \cdot 055\mu_B$ and that there is a small negative contribution $(-0 \cdot 105\mu_B)$ oppositely directed to the 3d spin and associated with the s electrons [8].

If we discount the alarming idea that fractions of electrons could exist, the explanation of these phenomena requires that each electron (or some of the electrons) cease to be rigidly associated with one particular ion. This is suggested by the high electrical conductivity of metals, which indicates considerable freedom of motion of the electrons through the crystal lattice.

Table 4.3.

	I_s	I_o	n_B	T_C °C[a]	$K_1 \times 10^5$
Fe	1710	1752	2·22	770	4·8
Co	1431	1446	1·71	1120	53
Ni	485	510[b]	0·606[b]	358	0·45
Gd		1950	7·55[c]	20	
Tb		$\sigma_o = 328$	9·34	−52[d]	

[a] This is the ferromagnetic Curie point. The paramagnetic Curie temperatures, given by $\chi = C/(T - T_C)$, are a little higher.

[b] Recent measurements by Danan et al. [30] give $\sigma_o = 58 \cdot 6$ e.m.u. g^{-1}, while the previously accepted value, which corresponds to I_o in the table, was $\sigma_o = 57 \cdot 5$ e.m.u. g^{-1}. This correction is extremely important since pure nickel is often used for calibrating magnetometers.

[c] Expected value for atomic gadolinium is $7 \cdot 0$ ($M_S = \frac{7}{2}$, $L = 0$).

[d] Terbium becomes antiferromagnetic above 221°K and paramagnetic above 229°K.

Note also that some alloys of other metals, notably manganese (MnBi, MnSb) are ferromagnetic.

4.3.1 Free electron theory

This rather extreme type of theory treats the outer electrons as moving with complete freedom throughout the crystal, bound only by the limits of the crystal itself. These outer electrons can reasonably be taken to be those which are readily lost by the isolated atoms to form ions of the most common valency. For example, sodium can be considered to give one free electron or *conduction electron* per atom $(3s^1)$ leaving a localized neon core, that is a structure very similar to the neutral neon atom but having, of course, a single positive charge. Potassium can similarly be represented as argon, $4s^1$, and aluminium as neon, $3s^23p^1$, with three conduction electrons per atom.

The positive ion cores are initially left completely out of account and are supposed not to affect the motion of the free electrons. They make a diamagnetic contribution to the susceptibility, which must be used as a correction to the magnetic effects of the free electrons.

Having postulated that the free electrons are to be treated collectively, it then follows, according to the exclusion principle, that they must each occupy a different energy level, or occupy kinetic levels in pairs with $m_s = \pm\frac{1}{2}$. Consequently, there must be a great number of energy levels with only small spacing between them. At absolute zero the electrons will fill the lowest of the levels in this nearly continuous band. The population of electrons continues up to an energy level which depends upon the total number available and on the distribution of the energy levels. This is called the *Fermi level*.

Thus, the simple representation of a free electron system is as shown in Figure 4.11. E_s is the surface potential, which can be measured by slow electron diffraction, and in the case of nickel is $14\cdot8$ eV. ϕ is the work function, which gives the energy needed to remove an electron from the Fermi level right out of the metal; this can be measured by experiments on thermal emission, photoemission, etc. The diagram is really based on the simple observation that electrons do not leave the metal spontaneously. The full description of the system requires the derivation of the energy levels and, fortunately, the wave-mechanical calculation is very simple, particularly if the one-dimensional case is taken.

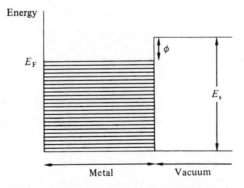

Figure 4.11. A diagrammatic illustration of closely spaced energy levels in a metal crystal, according to the free electron theory. E_F is the highest occupied level (Fermi level), E_s the surface potential and ϕ the work function.

There is no potential energy for the model, so the Hamiltonian represents the kinetic energy only. The operator for linear momentum is

$$\hat{p} = -i\hbar \frac{\partial}{\partial x}. \tag{4.39}$$

By the correspondence principle the kinetic energy operator is $\hat{p}^2/(2m)$ (since classically $p = mv$, kinetic energy $= \frac{1}{2}mv^2$) and so

$$\hat{H} = \frac{(-i\hbar)^2}{2m} \frac{\partial^2}{\partial x^2}, \tag{4.40}$$

and the Schrödinger equation,

$$\hat{H}\psi_n = E_n\psi_n,$$

becomes

$$-\frac{\hbar^2}{2m} \frac{d^2\psi}{dx^2} = E\psi.$$

This equation can be rewritten as

$$(D^2 + k^2)\psi = 0, \text{ where } k^2 = (2m/\hbar^2)E,$$

with the general solution

$$\psi = Ae^{ikx} + Be^{-ikx}$$

in which A and B are constants of integration. The boundary conditions are $\psi = 0$ for $x = 0$ and $x = L$, where L is the length of the crystal parallel to OX. The first condition gives

$$0 = A \times 1 + B \times 1$$

$$\psi = A(e^{ikx} - e^{-ikx}) = A\sin kx.$$

The second condition gives

$$\sin kL = 0, \text{ whence } kL = n\pi.$$

Hence the eigenfunctions are

$$\psi_n = A_n \sin(n\pi x/L),$$

and the energy levels are

$$E_n = \frac{n^2\pi^2\hbar^2}{2mL^2} \qquad n = \pm 1, \pm 2, \ldots$$

The three-dimensional Schrödinger equation is given by replacing $\partial^2/\partial x^2$ by ∇^2: it can be divided into three identical ordinary differential equations on the assumption that $\psi(x, y, z)$ is the product of three separate functions of x, y, z, and so the final solution is represented by the eigenfunction

$$\psi(x, y, z) = C\sin(n_x\pi x/L)\sin(n_y\pi y/L)\sin(n_z\pi z/L). \tag{4.41}$$

This function represents standing waves, the nodes and antinodes being stationary. The energy levels are now given by the eigenvalues

$$E_n = \frac{\hbar^2 k_n^2}{2m} = \frac{\hbar^2 \pi^2 n^2}{2mL^2},\tag{4.42}$$

(since the boundary conditions now give $k = 2\pi n/L$), where

$$n^2 = n_x^2 + n_y^2 + n_z^2.\tag{4.43}$$

Note that there are now a number of combinations of n_x, n_y and n_z which give a particular value of n and thus of the energy. If we define a 'state' by a certain set of values of these quantum numbers, it is clear that each energy level is degenerate, and that the degeneracy increases as the energy increases. Furthermore the spacing between the energy levels, when $n \rightarrow (n+1)$, decreases as n increases. (The spacing is always very small: $\hbar^2 \pi^2/2mL^2 = 10^{-16}$ eV for $L = 1$ cm, so that the energy levels form a nearly continuous band for all crystals of substantial size.) Even for $n = 1$ there are six states (neglecting the spin) which may be written with an obvious symbolism as

$$n^2 = \begin{cases} 1^2 +0+0, & 0+ 1^2 +0, & 0+0+ 1^2, \\ (-1)^2+0+0, & 0+(-1)^2+0, & 0+0+(-1)^2. \end{cases}$$

Remembering that each electron can have two values of m_s, namely $\pm\frac{1}{2}$, in addition to the quantum numbers giving the kinetic energy, we may say that either there are 12 states for $n = 1$ or that there are 6 states which can be doubly occupied. The problem now is to determine just how the density of states varies with the energy and to find the Fermi level. At the absolute zero of temperature this 'density of states curve' will also represent the energy distribution of the electrons, the Fermi energy being that of the most energetic electrons; at finite temperatures the manner in which the electrons are distributed among the available states and energy levels must be evaluated.

The magnitude of any vector $\mathbf{R}$ is $(R_x^2 + R_y^2 + R_z^2)^{\frac{1}{2}}$ where R_x, R_y and R_z are the components on rectangular axes. Thus, if we consider n as a vector, n_x, n_y and n_z can be treated as its components for the sake of computation, simply because of the relation which their squares have to n^2. However, in this particular case the components can have only integral values and so in two dimensions the space must be considered to be divided up into a net as shown in Figure 4.12. Values of $n^2 \doteqdot 8$ are given by all those combinations of n_x and n_y indicated by the points, and values of $n^2 < 8$ are given by the pairs of values of n_x and n_y which fall within the circle. It is clear that as the value of n becomes great the number of combinations of n_x and n_y for $n^2 < n'^2$ approaches the area of the circle (the net having unit spacing and one point per unit cell). In three dimensions the number of combinations for $n^2 < n'^2$ is obviously given by the volume of the sphere of radius n', or $\frac{4}{3}\pi n'^3$.

This value must be multiplied by 2 to give the number of states including the effect of spin: if this factor is omitted, it must be assumed that two electrons ($m_s = \pm\tfrac{1}{2}$) can occupy each state.

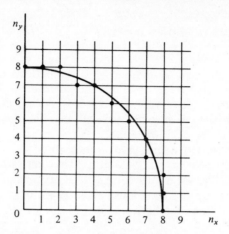

Figure 4.12. A two-dimensional square net with unit spacing and cells of unit area, showing the range of values of n_x and n_y which give a certain value of n: $n \doteq 8$ for all the points shown and $n < 8$ for all the points within the circle (or sphere, in three dimensions).

Suppose the electron concentration is N per unit volume; the total number of electrons is NL^3 and at absolute zero all will have energies below the Fermi level

$$E_F = \frac{2\pi^2\hbar^2}{mL^2}n_F^2. \tag{4.44}$$

Thus, accounting for spin

$$NL^3 = \tfrac{8}{3}\pi n_F^3.$$

and this equation gives the value of n_F from which

$$E_F = \frac{2\pi^2\hbar^2}{mL^2}\left(\frac{3NL^3}{8\pi}\right)^{2/3} = \frac{\hbar^2}{2m}(3\pi^2 N)^{2/3}. \tag{4.45}$$

N is taken as the number of valence electrons. Calculated values of E_F, and the Fermi temperature given by $kT_F = E_F$, are $3\cdot1, 2\cdot1, 1\cdot5$ eV and $37\,000, 24\,000, 18\,000°$K respectively for sodium, potassium and caesium: they are quite close to the values indicated by experiment.

The density of states curve itself is given by the distribution of energies at absolute zero. Suppose $Z(E)$ is the number of states with energy E; then for unit volume the integral of $Z(E)$, up to a value of E which corresponds to a certain value of n, is equal to the number of states included by the corresponding sphere, that is

$$\int Z(E)\mathrm{d}E = \tfrac{4}{3}\pi n^3 = \tfrac{4}{3}\pi\left(\frac{mE}{2\pi^2\hbar^2}\right)^{3/2}$$

after substituting for n. Differentiating with respect to E gives

$$Z(E) = (\text{constant})E^{\frac{1}{2}}, \qquad (4.46)$$

which is a parabola as shown in Figure 4.13a.

At absolute zero the distribution of electron energies $N(E)$ is identical with the density of states curve, cut off at the Fermi level. [$N(E)$ is the number of electrons with energy E.] At finite temperatures $N(E)$ is obtained from $Z(E)$, the number of available states, by multiplying the latter by the function which represents the distribution of the electrons over the states. Specifically $N(E) = F(E)Z(E)$ where $F(E)$ is the Fermi function. This is derived, taking as our principle that electrons obey Fermi–Dirac statistics, as

$$F(E) = \frac{1}{\exp\left[(E-E_{\mathrm{F}})/kT\right]+1} \cdot \qquad (4.47)$$

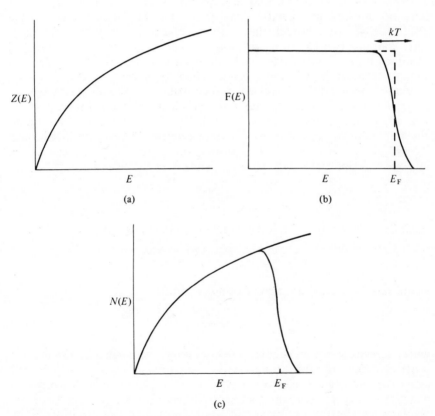

Figure 4.13. (a) Parabolic density of states curve: $Z(E) \propto E^{\frac{1}{2}}$. (b) The Fermi function $F(E)$, which indicates the occupation of the levels whose density is given by $Z(E)$. The broken line applies to absolute zero. (c) The distribution of electron energies for a free electron gas, that is $N(E) = F(E) \times Z(E)$.

We will not reproduce the standard statistical treatment, but note that as $T \to 0$, $F(E) \to 1$ for all values of $E < E_F$ and $F(E) \to 0$ for values of $E > E_F$. Thus E_F corresponds to the description already given and $F(E)$ only has the effect of cutting off $N(E)$ at this level. At finite temperatures $F(E)$ has the form shown in Figure 4.13b, a step function which is 'smeared out' over a range of approximately kT. Thus $N(E)$, in this case, has the form shown in Figure 4.13c.

4.3.2 Pauli paramagnetism

It is recalled that certain metals have low paramagnetic susceptibilities which are nearly independent of temperature. This behaviour is certainly not explained by the theories already given, which lead to $\chi = C/T$ or $\chi = C/(T+\theta)$; it is, however, explained by the free electron theory [9].

It must be remembered that in a quantized system magnetization can only be induced by specific changes in the spin components of the electrons. If we consider the states representing the kinetic energies, each contains two electrons with $m_s = \frac{1}{2}$ and $m_s = -\frac{1}{2}$, and the reversal of either spin would violate the Pauli exclusion principle if the electron were to remain in the same level. Thus, only the spins of electrons in unfilled levels can be reversed, and contribute to the induced magnetization. (Similar principles apply to the electronic contribution to the heat capacities of metals; these are $\sim 0\cdot01$ of the classical values due to the impossibility of thermally exciting electrons between filled levels.) According to Figure 4.13c the proportion of the electrons in unfilled levels is very approximately kT/E_F. Thus N in the equation

$$\chi = \frac{N\mu_B^2}{kT}$$

must be replaced by the effective value of NkT/E_F, giving

$$\chi = \frac{N\mu_B^2}{kT} \times \frac{kT}{E_F} = \frac{N\mu_B^2}{E_F} = \frac{N\mu_B^2}{kT_F}.$$

A full statistical treatment gives a similar result:

$$\chi = \frac{3N\mu_B^2}{2kT_F}, \tag{4.48}$$

which is clearly independent of temperature and is also of the correct order of magnitude for lithium, sodium and potassium, namely $\chi \doteqdot 10^{-6}$ assuming one valence electron per atom ($N \approx 10^{22}$ cm^{-3}). In order to test the theory experimental susceptibilities must be corrected for the diamagnetism of the ion cores and also for a diamagnetic effect from the conduction electrons themselves. As an approximation the spin and orbital motion of the electrons can be treated separately: the spin as above and the orbital motion to give initially a calculated diamagnetic susceptibility which, rather remarkably, is just one-third of the para-magnetic effect of the spins, that is

$$\chi = -\frac{N\mu_B^2}{2kT}.$$
(4.49)

However, this expression must again be corrected to apply to electrons near the Fermi level only, which reduces the magnitude and removes the temperature dependence. (The formulae for both the paramagnetic and diamagnetic susceptibilities indicate a *slight* temperature dependence, since neither E_F nor T_F is not quite constant.)

From the experimental point of view it is important to note that the paramagnetic spin susceptibility itself can be isolated from the other effects by resonance techniques [10, 11], specifically by comparing the areas under the absorption curves obtained for electron spin resonance and for nuclear magnetic resonance. This method gives $\chi = 2\cdot08 \times 10^{-6}$ and $0\cdot95 \times 10^{-6}$ for lithium and sodium respectively, for which the Pauli formula gives $1\cdot17 \times 10^{-6}$ and $0\cdot64 \times 10^{-6}$. Considerably better agreement is obtained by taking some account of the interactions of the electrons with the ion cores, which can be done by replacing the actual mass of the electron m by an *effective mass* m^*: the expression $E = \hbar^2 k^2/2m$ is replaced by

$$E = \frac{\hbar^2 k^2}{2m^*},$$
(4.50)

and it is shown in the standard works that m^* can be either greater or less than m. The best agreement with the experimental results is obtained when the Pauli formula is also modified to take account of interactions (exchange and correlation) between the electrons [12].

In the face of these forbidding complications it is worth emphasizing that the remarkably simple theory of perfectly free electron behaviour gives quite a satisfactory account of the magnetic properties of some metals. On the other hand, there are clearly many metals to which it cannot apply at all directly.

4.3.3 Localized moments in metals

The free electron theory completely fails to explain the occurrence of high, temperature-dependent, magnetic susceptibilities or the spon-taneous ferromagnetism and antiferromagnetism in the 3d metals. As a deliberate diversion one might introduce the following scheme:

(a) make a clear distinction between the inert argon core (henceforth neglected), the 3d electrons as in the divalent ions, and the valence electrons (mainly 4s);

(b) consider the valence electrons to be completely free, giving Pauli paramagnetism, and

(c) consider the 3d electrons as localized and non-interacting in order to give local moments as in ionic crystals. With the orbital moment quenched this would give $S = 0 \to \frac{5}{2} \to 0$ and components in μ_B of $0 \to 5 \to 0$ for the series $3d^0$ to $3d^{10}$ (Ca^{2+} to Zn^{2+}).

However this scheme is clearly invalidated by the observation that the real atomic moments are much smaller than those forecast, and are not even integral. The moments can be determined from the saturation magnetization (with crystallographic data) for the ferromagnetic metals, and from neutron diffraction studies for either ferromagnetics or anti-ferromagnetics. The moments for chromium, iron, cobalt and nickel, in the magnetically ordered state, are approximately 0·4, 2·2, 1·6 and $0·6\mu_B$ compared with 4, 4, 3 and $2\mu_B$ for the free ions.

Neutron diffraction, which is capable of giving spin-distribution maps similar to the electron density maps obtained by X-ray diffraction, admittedly indicates that the moments are localized to a certain extent in the 3d metals. For nickel [8], Figure 4.14 indicates that 80% of the 'magnetic electrons' are in t_{2g} orbitals, and the authors showed that the magnetization could be divided into three contributions: $0·65\mu_B$ from the 3d spin, only $0·055\mu_B$ from the 3d orbit which is very substantially quenched, and a negative contribution of $-0·105\mu_B$. This latter value may reasonably be associated with the 4s electrons, implying that they are negatively exchange polarised with the localized moments.

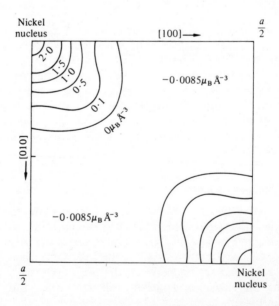

Figure 4.14. The distribution of the magnetic moment (spin density) in the (100) plane of a nickel crystal, as revealed by the diffraction of polarized neutrons (Mook and Shull [8]).

In the 4f, rare earth, metals the unpaired spins in the 4f shell are shown by neutron diffraction to occupy a very small proportion of the space occupied by the atoms. Also, the number of unpaired spins for the metal is, in some cases at least, very similar to that for the free atom or ion in a compound. For example gadolinium is ferromagnetic below room temperature (Curie point 16°C) and has a moment of nearly $7\mu_B$, corresponding to the seven unpaired electrons. (The orbital moment is not expected to be quenched in the 4f metals, but $L = 0$ in this case anyway.) Generally, then, the simple picture of localized moments may apply to these metals, but not to the 3d series.

4.3.4 Simple band theory

The free electron theory can be considered as a rudimentary band theory, in which all the valence electrons occupy a single hyperbolic band, and all the others are completely neglected. For example, Pauli paramagnetism can be represented pictorially by Figure 4.15. In the absence of a field the band may be divided into two identical sub-bands, since each kinetic state can be considered as the superposition of the states for $m_s = \pm\frac{1}{2}$; one sub-band contains all the electrons with $m_s = +\frac{1}{2}$ and the other with $m_s = -\frac{1}{2}$. The signs are initially quite arbitrary, but when a magnetic field is applied the positive sign can be taken to apply to those electrons which have a spin component parallel to the field direction and vice versa. Thus, the effect of the field, H, can be represented by shifting the bands relative to each other, through a distance on the energy axis of $\Delta E = H\mu_B$ (Figure 4.15b). Finally, the electrons are considered to 'flow over' into the lower band to equalize the levels (Figure 4.15c).

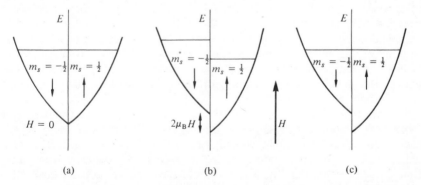

Figure 4.15. Representation of Pauli paramagnetism. (a) Two parabolic bands containing the electrons with $m_s = +\frac{1}{2}$ and $m_s = -\frac{1}{2}$ separately, each being given by $\frac{1}{2}N(E)$ in the absence of a magnetic field. (b) The shift of the bands due to an applied field, H, assuming that $m_s = +\frac{1}{2}$ represents those electrons with spin components parallel to the field. (c) The equilibrium state achieved in an applied field.

There are two different ways of showing how a more complex band structure can arise. The first is to modify the Hamiltonian for the free electron by introducing a term for the potential energy of the electrons, the potential function being periodic with the same period as that of the lattice, that is $V(\mathbf{r}) = V(\mathbf{r+d})$ where $\mathbf{d}$ is the lattice vector. The solutions of the Schrödinger equation, with Hamiltonian $(\hat{E} + \hat{V})$, are found to have the form

$$\psi_k = U_k(\mathbf{r})e^{i\mathbf{k} \cdot \mathbf{r}}, \tag{4.51}$$

where $\mathbf{k}$ is the wave vector whose magnitude is the reciprocal of the wavelength and which is pointed in the direction of motion of the plane waves. Also, $U_k(\mathbf{r}) = U_k(\mathbf{r+d})$, that is it has the periodicity of the lattice. The wavefunctions ψ_k are known as *Bloch* functions, and evaluation of their eigenvalues shows that these are no longer quasi-continuous as for free electrons, but can only fall in certain ranges of energy, or bands.

A more 'visual' approach commences at the opposite extreme by considering the effects of bringing initially isolated atoms into close proximity (tight binding approximation). For example, if two hydrogen atoms have wavefunctions ψ_A and ψ_B, the wavefunction for the pair is

$$\psi = \psi_A \pm \psi_B. \tag{4.52}$$

The positive sign, which gives a symmetric spatial wavefunction, anti-parallel spins and a high electron density between the nuclei, corresponds to the lowest energy, that is to bonding as discussed in section 4.2 on direct exchange. When the spacing, d, is very great the two alternatives must give the same energy, so the energy of the pair is expected to vary with d as shown in Figure 4.16a. Calculations for a hypothetical crystal consisting of six hydrogen atoms, again made by the 'linear combination of atomic orbitals' method, give the results shown schematically in Figure 4.16b. Drawing a line at the equilibrium separation gives the expected energy levels shown in Figure 4.16c, the lines on the left corresponding to the Bohr energies calculated for the isolated hydrogen atoms. In this case there are clearly 'forbidden bands' of considerable width; the diagram is not to scale and the spacing of the levels is much smaller than the width of the gaps.

In the ground state, at $0°K$, the six electrons of the hypothetical hydrogen crystal occupy only the three lowest levels, since $m_s = \pm\frac{1}{2}$ for each. Thus, the lowest band is only half full, and on the application of a magnetic field the spin states can change freely. However, if we assume that a hypothetical helium crystal would have the same band structure, the lower band would be completely filled by the 12 electrons, again in the ground state. On the further assumption that the gap between this and the second lowest band is greater than $\mu_B H$ for any attainable field, this latter model is diamagnetic in nature (and is also an insulator, since the kinetic energy cannot change in a full band). The hydrogen model, on the other hand, exhibits paramagnetism.

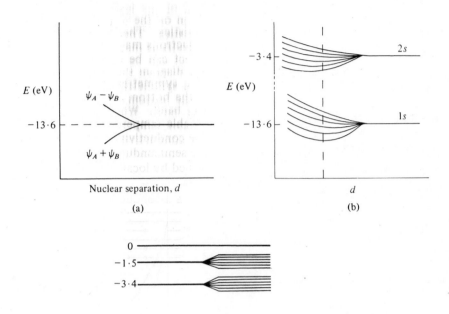

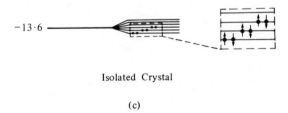

E (eV)

Isolated Crystal

(c)

Figure 4.16. (a) The change in the energy of a pair of hydrogen atoms as the internuclear separation, d, is changed. On the right of the diagram it is assumed that d is so great that the interaction is negligible.

(b) Schematic representation of energy levels for a hypothetical crystal of six hydrogen atoms.

(c) The occupation of the energy levels given by (b) in the ground state (absolute zero).

At finite temperatures the occupation of the levels in the bands is given by superimposing the Fermi function on the band structure, since the electrons still obey Fermi–Dirac statistics. Then, if the spacing of the bands is of the order of kT, some electrons may be excited from a lower, initially filled, band and the effect can be represented for the helium model as in Figure 4.17. In this diagram the Fermi function is superimposed on the band structure in a symmetrical manner, since it can be shown that the value of $F(E)$ at the bottom of the upper band is equal to $1 - F(E)$ at the top of the lower band. When electrons can be thermally excited across a gap at reasonable temperatures, the material exhibits semiconductivity with increasing conductivity at increasing temperatures. Most of the spinel ferrites are semiconductors, although their magnetic properties are adequately described by localized models.

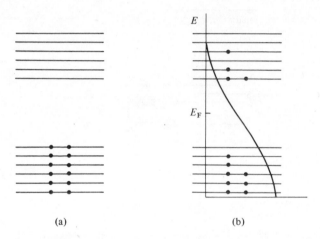

(a)   (b)

Figure 4.17. (a) A full band at absolute zero, as typified by a hypothetical helium crystal with the same energy levels as the hydrogen model but twice as many electrons.
  (b) As for (a) but at a temperature for which kT is greater than the energy gap. The model now represents semiconductivity and the magnetic properties will also be modified.

4.3.5 Overlapping bands

In the crude models outlined above the resulting bands are separate, but when more realistic calculations are made, for example for the 3d metals, it is found that the bands overlap and the separate identity of the 3d and 4s electrons is lost.

It seems impracticable to attempt even to survey the relevant methods of calculating the band structure of real metals: the tight-binding method indicated above is easy to envisage, although it does not work very well in practice and other more useful methods are described reasonably simply by Ziman for example [13]. Schematically, the results of the calculations for 3d metals can be represented as in Figure 4.18, and according to the rigid band approximation this can be taken to apply to all the 3d series. The difference between the elements corresponds to the number of 3d and 4s electrons per atom and thus the levels to which the bands are filled—the Fermi levels.

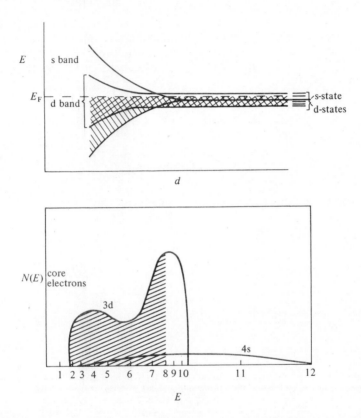

Figure 4.18. (a) The combination of single-atom s-states and d-states to form bands which overlap in energy. Note the wide spread of the s band to include states with very high energy and equivalent velocity (after Ziman [13]).

(b) The approximate density of states curve corresponding to (a), application to different members of the 3d series depending on the numbers of electrons per atom and thus the position of the Fermi level.

The important point to note about Figure 4.18 is that the s band spreads over a wider range of energies than the d band. As well as having a narrow spread at the equilibrium lattice spacing, the 3d band contains five times as many states as the 4s band. Thus, when the number of states is plotted against their energy, it is clear that the resulting density of states curve must look something like Figure 4.18b, but the details of the shape of the d band remain uncertain.

Accepting that the density of states curves can be calculated, and checked by a variety of methods, some obvious points may be made with regard to their magnetic significance. Firstly, if the Fermi level is above the top of the d band, the electrons in this band cannot make any contribution to the magnetic susceptibility or to the spontaneous magnetization. Secondly, since only the electrons at the Fermi surface contribute to the susceptibility, the magnitude of the susceptibility must depend on the density of states at the Fermi level. To be more specific, (referring back to Figure 4.15), the total number of electrons in the + spin states, n_+, is

$$n_+ = \int_0^\infty \tfrac{1}{2} N(E) F(E - \mu_B H) \, dE ,$$

and similarly

$$n_- = \int_0^\infty \tfrac{1}{2} N(E) F(E + \mu_B H) \, dE .$$

With the orbital moment quenched, the magnetization is

$$I = \mu_B (n_+ - n_-)$$

$$= \mu_B \int_0^\infty \tfrac{1}{2} N(E) [F(E - \mu_B H) - F(E + \mu_B H)] \, dE$$

$$\doteqdot \mu_B^2 H \int_0^\infty -\left(\frac{\partial F}{\partial E}\right) N(E) \, dE$$

$$= \mu_B^2 H N(E_F) ,$$

whence

$$\chi = \mu_B^2 N(E_F) \qquad\qquad (4.53)$$

is directly proportional to the density of states at the Fermi level. Consequently, a large number of unpaired spins, for a localized moment model, is equivalent to a high density of states at the Fermi level for the band model. On looking again at Figure 4.18 it is clear that the contribution of the 4s electrons to the susceptibility can never be very great, wherever the Fermi level, and it is the peak in the 3d band which ensures that, qualitatively at least, the band theory is successful in explaining high magnetic susceptibilities. The values for b.c.c. alloys and f.c.c. elements given in Figure 4.19 were directly associated with the band structure by Taniguchi, Tebble and Williams [14]. It has also been found that the extremely high susceptibility of palladium may be reduced progressively towards zero when the metal absorbs hydrogen, the electrons donated by the hydrogen appearing to fill the 4d band.

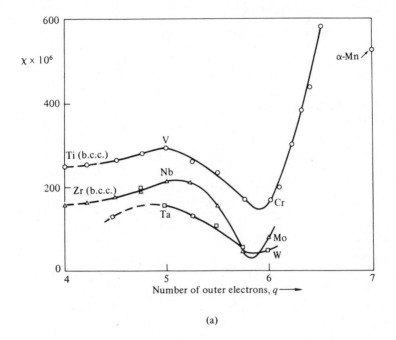

(a)

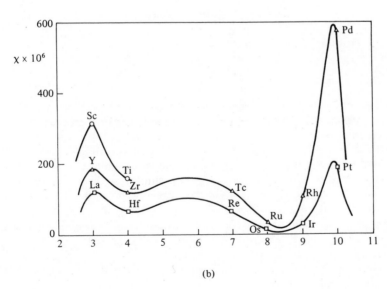

(b)

Figure 4.19. (a) Susceptibility of b.c.c. alloys at room temperature (Taniguchi *et al.* [14]).
(b) Susceptibility of f.c.c. metals at room temperature (Taniguchi *et al.* [14]).

4.3.6 Ferromagnetism and band theory

The application of the band theory to ferromagnetism, as first carried out by Stoner [15], requires the introduction of an exchange term into the treatment given above. The exchange energy must be proportional to the magnetization, that is proportional to the mean resolved moment of the spins, $\langle \mu \rangle$. If an interaction parameter θ' is used, such that the exchange energy for the electrons is

$$E_{ex} = \pm \frac{k\theta'\langle \mu \rangle}{\mu_B} , \qquad (4.54)$$

where k is Boltzmann's constant, the energies of the spins parallel and antiparallel to an applied field E_+ and E_-, are

$$E_+ = E_0 - \mu_B H - \frac{k\theta'\langle \mu \rangle}{\mu_B}$$

and

$$E_- = E_0 + \mu_B H + \frac{k\theta'\langle \mu \rangle}{\mu_B} ,$$

bearing in mind that the spins with $m_s = +\frac{1}{2}$ are those which have a moment parallel to the field and thus to the net magnetization. As in the paramagnetic case, the magnetization is

$$I = n\langle \mu \rangle = \mu_B \int_0^\infty \tfrac{1}{2}[F(E_+) - F(E_-)]N(E)\,dE . \qquad (4.55)$$

The total number of electrons is given by

$$n = \int_0^\infty \tfrac{1}{2}[F(E_+) + F(E_-)]N(E)\,dE . \qquad (4.56)$$

These two equations can be solved to find $\langle \mu \rangle$, and thus the magnetization, as a function of temperature. It is found that for certain values of the exchange parameter θ' the magnetization is finite even when $H = 0$, or in other words ferromagnetism occurs, while for smaller values the effect of the interaction is to increase the paramagnetic susceptibility. Thus, the results are qualitatively very similar to those obtained in terms of an exchange field or Weiss molecular field applied to localized moments, as outlined in Chapter 2, although the situation is so very different: in the band theory the exchange is that between individual electrons in the band as compared with exchange between the ionic moments as a whole (described in section 4.2).

Apart from the exchange parameter itself, the shape of the bands is also expected to influence the occurrence of ferromagnetism. If the band is parabolic in shape [$N(E) \propto E^{1/2}$)], it has been shown by Stoner [15] that spontaneous magnetism should occur for $k\theta'/E_0 > \frac{2}{3}$, where E_0 is the Fermi level at absolute zero. Calculations of the reduced spontaneous magnetization, and of the inverse susceptibility above the Curie point for a parabolic band, are shown in Figure 4.20, the variable parameter being $k\theta'/E_0$. Reference should also be made to work by Wohlfarth [16] on the modifications due to the overlap of the d and s bands.

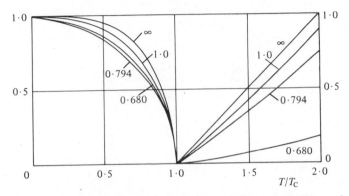

Figure 4.20. The calculated temperature dependence of the magnetization, and of the reciprocal of the susceptibility, above the Curie point in arbitrary units for a parabolic band. The numbers represent the values of $k\theta'/E_0$ (Stoner [15]).

A pictorial representation of the effect of exchange on the distribution of the spin directions, and thus on the magnetization, can be made as for Pauli paramagnetism but with a much bigger shift of the half bands, since the exchange coupling is equivalent to a field of about 10^7 Oe. Figure 4.21a represents the ferromagnetic band structure of nickel, the 'positive' d half band being full while the 'negative' d half band contains approximately 4·4 electrons and leaves 0·6 holes per atom, or 0·6 μ_B. There are also 0·6 4s electrons per atom with some small effect on the magnetization: whether the 4s half bands should be taken to be shifted, by how much and even in which direction is still a subject for speculation. An obvious symbolism for nickel is $3d^5\uparrow\, 3d^{4\cdot4}\downarrow\, 4s^{0\cdot6}$.

A similar picture applies to cobalt. With one less electron per atom one 3d half band is still full, and the configuration is $3d^5\uparrow\, 3d^{3\cdot4}\downarrow\, 4s^{0\cdot6}$. It might be taken to follow that iron would be $3d^5\uparrow\, 3d^{2\cdot8}\downarrow$, but it has been suggested by Mott [17] that in fact neither half band is full: the configuration is $3d^{4\cdot7}\uparrow\, 3d^{2\cdot4}\downarrow\, 4s^{0\cdot9}$ as represented by Figure 4.21b.

4.3.7 Localized moments and band theory: exchange

The band theory, or Stoner collective electron theory, treats electrons as belonging to the metal crystal as a whole and having no correlation with the ion cores, while the Weiss theory treats electrons as belonging to specific ions. The latter is conceptually simple but fails to explain the non-integral moments. Further, the theory of exchange developed for the hydrogen molecule in terms of the direct overlap of orbitals only

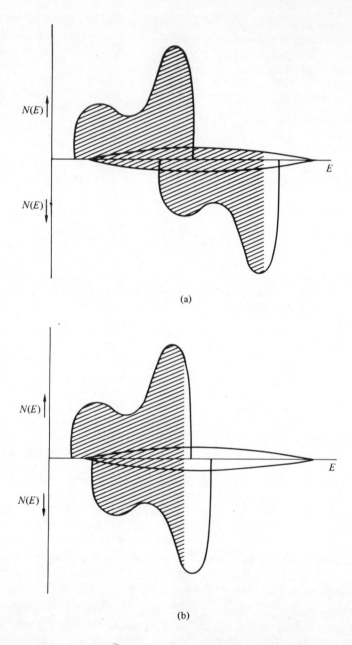

Figure 4.21. (a) Schematic representation of the band structure of nickel in the ferro-magnetic state, with 0·6 holes in the negative half band. (b) Ferromagnetic band structure for iron, with neither half band full.

applies at all directly to the Weiss model. Figure 4.22, which shows a very interesting empirical relation between the sign (and magnitude?) of the exchange integral and the lattice spacing, could be reasonably interpreted in terms of the amount of overlap. The results of neutron diffraction also appear to indicate a considerable localization of the 'magnetic' electrons in 3d metals.

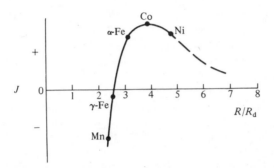

Figure 4.22. A representation of the exchange interaction as a function of interatomic spacing in transition metal crystals. γ-iron and manganese are antiferromagnetic.

It may well be that the true picture lies between the extremes of these two models, the electrons being regarded as free inasmuch as they do certainly move through the lattice and contribute to the electrical conductivity but localized inasmuch as they reside in the vicinity of the ion cores for the majority of the time. While the electrons are close to the ion cores or lattice sites, their wavefunctions may be very similar to those for isolated 3d ions. Some experimental results (determinations of susceptibility and neutron scattering above the Curie temperature) indicate that iron is close to the Weiss extreme, while other data indicate that nickel and cobalt are closer to the collective electron extreme. However, it seems that calculations can only be made at all readily for the two extremes and not for intermediate cases (see also page 129).

It should be stressed that the above discussion has mainly concerned the 3d metals. The 4f metals, however, are almost certainly adequately represented by a localized model. The magnetic electrons in this case are very tightly bound in a small proportion of the ionic volume, the radius of the f shell being one-tenth the ionic spacing, and are shielded against the effects of the environment by the 5s and 5p electrons (cf. the relative ineffectiveness of crystal fields on the orbital momentum). In this case, however, it is impossible to derive the exchange interactions from direct overlap of the orbital wavefunctions. Kittel and others [18] have suggested that the exchange may be transmitted from one localized moment to another by the exchange polarization of the conduction electrons.

This suggestion appears to be rather similar to the mechanism put forward by Vonsovskii [19] for the 3d metals. In this latter proposal the primary effect was considered to be direct Heisenberg exchange between localized 3d shells, with polarization of the conduction electron spins parallel to the direction of the local moments so that they contributed to the magnetization (but see page 116). Zener [20], on the other hand, suggested that there was no need for the primary direct ionic exchange, but that the alignment would be produced by the polarization of the 4s electrons themselves. Stuart and Marshall [21] and Watson [22] have shown that the exchange calculated for direct overlap of the d orbitals is very much smaller than that required, and may even be negative for the ferromagnetic metals.

A reasonably simple account would seem to be that when an electron is in the vicinity of a nucleus, it has a wavefunction which is very similar to an ionic d wavefunction, but in between the nuclei the wavefunction must be similar to that for purely collective or quasi-free electrons (for example, see Ziman [13] Chapter 3). It does not appear to be necessary at first sight to discriminate between 4s conduction electrons as such, or between itinerant 3d electrons which spend most of their time near to the nuclei, but move from one ion to another. Neutron diffraction results, as already noted, indicate the d-nature of the ions themselves. Now while an electron is in the locality of an ion its spin must be governed by Hund's first rule, which controls the spin components of the ion itself: in the 3d metals the shells are more than half full and so the mobile electron should be polarized antiparallel to the ionic moment.

As this negative-spin electron moves through the lattice it must conserve its spin direction to a considerable extent. The reversal of the spin involves a change of energy and in consequence can only happen when some process occurs whereby this energy can be given a different form, for example by the creation of a phonon. The spin–lattice relaxation time for this process is amenable to direct study by resonance methods: spin-reversal may only occur at a small fraction of the collisions limiting the conductivity, particularly if the spin–orbit coupling is low. The conservation of the spin direction, for times considerably greater than those taken to move between lattice sites, carries the spin polarization from one site to another. It is noted that the neutron diffraction results for nickel indicated that the 4s electrons made a negative contribution to the magnetization (section 4.3.3).

4.3.8 Alloys

The variation of the spontaneous magnetization of nickel, when it is alloyed with non-magnetic metals, is quite different from that of iron.

The magnetization of solid solutions of nickel and copper falls linearly to zero at 60% copper, as shown in Figure 4.23. If we treat the electrons of the two elements collectively and assume a constant band structure, this result is readily interpreted as the progressive filling of the spin-down half band which initially contains 0·6 holes per atom, as indicated. The effect of alloying with other elements may be similarly interpreted, and represents very powerful evidence for the collective electron theory of magnetism.

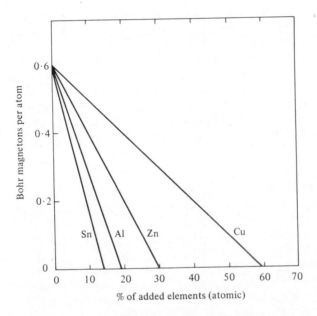

Figure 4.23. Spontaneous magnetization of nickel alloys forming continuous solid solutions. These are simply interpreted as representing the filling of the 0·6 holes in the 3d band by the one extra valence electron per atom of Cu, with 2, 3 and 4 valence electrons for Zn, Al and Sn respectively (from Bozorth [28]).

However, when iron is alloyed with many other non-magnetic metals, the effect is simply one of dilution. There is a decrease in the magnetization of very nearly one atomic moment for iron for each non-magnetic atom dissolved, as if there were no interaction at all. The magnetization approaches zero only when the concentration of iron approaches zero.

Some dilute alloys of iron have a remarkably high magnetization, and it appears that the iron can exchange-polarize the surrounding paramagnetic ions to give huge effective moments. This is a particularly current field of research, and many references will readily be found in the recent literature (for example, by recourse to the Conference Supplements in the Journal of Applied Physics).

4.4 ANISOTROPY AND ORBITAL ANGULAR MOMENTUM

Although magnetocrystalline anisotropy is obviously connected with the crystalline environment, no direct interaction between the lattice and the electron spin momenta appears to have been postulated. The interactions are with the orbital angular momenta, an effective interaction with the spin following from the spin–orbit coupling. Thus, it seems reasonable to precede the discussion of anisotropy itself by some notes on orbital momenta, particularly since so much of the general theory of magnetic behaviour is devoted, naturally, to the electron spin.

Among the 3d transition ions the cobaltous ion has the greatest angular momentum, and this is the ion which gives rise to the highest anisotropy; for example, a small percentage of cobalt in a spinel ferrite can dominate the overall anisotropy, overcoming the effects of the other cations present. Since some selection must clearly be made, the following sections will deal mainly with the cobaltous ion, commencing with a brief review of cobalt oxides, which illustrate many interesting features.

4.4.1 Oxides of cobalt

Cobalt forms two single oxides and many mixed oxides with other 3d metals. CoO has a Néel point of $291°K$ and Co_3O_4 is also antiferromagnetic below $40°K$. By analogy with the spinel ferrites, it might be expected that Co_3O_4 would be ferrimagnetic below the Néel point, since it has a similar spinel crystal structure with Co^{3+} in octahedral sites and Co^{2+} in tetrahedral sites. However, the structure is slightly modified and the Co^{2+} ions are antiferromagnetically ordered on two different kinds of tetrahedral sites [23], while the Co^{3+} ions are in the 'low-spin' state and contribute no moment, however they are arranged. Isomorphic aluminites, $CoAl_2O_3$, $MnAl_2O_3$ and $FeAl_2O_3$ are also known, and Co_3O_4 has the highest Néel point of this group in spite of the diamagnetic nature of the Co^{3+} ions.

The ionic moment of Co^{2+} in CoO is $5·10\mu_B$. Magnetic cobaltic ions occur in the mixed oxide $Li_{0·5}Co_{0·5}O$, which can be considered as

$$4Li_{0·5}Co_{0·5}O = Li_2OCo_2O_3 ,$$

since lithium is monovalent. In oxides containing less lithium both Co^{2+} and Co^{3+} ions occur (for example $8Li_{0·25}Co_{0·75}O = Li_2O4CoOCo_2O_3$). The cobaltous ions then have a moment of $5·15\mu_B$ and the cobaltic ions $2·1\mu_B$.

$LiCoO_2$ has the strange property of being diamagnetic at low temperatures and paramagnetic at higher temperatures [24], the cobaltic ions changing from low-spin to high-spin states. $LaCoO_3$ is a perovskite which has a similar behaviour.

There are several very weakly magnetic spinel oxides which contain cobalt, such as nickel cobaltite, $NiCo_2O_4$, cobalt chromate and vanadate, $CoCr_2O_4$ and CoV_2O_4, and cobalt manganites: as a typical example, $Co_{1.1}Mn_{1.9}O_4$ has a moment of $0.5\mu_B$ per formula unit resulting from the canting of the ionic moments in the otherwise antiferromagnetic structure.

Another interesting phenomenon is exhibited by cobalt manganates, $CoMnO_3$, with the perovskite crystal structure. The magnetic moments of the cobalt ions are all ordered in the same direction below the Néel point, and those of the manganese ions are antiparallel to this direction. Two possible combinations satisfy the formula: Co^{2+} with Mn^{4+} or Co^{3+} with Mn^{3+}. Co^{2+} is $3d^7$ and thus has $(5-2)$ or 3 unpaired spins; Mn^{4+} is $3d^3$ with 3 spins. Also, Co^{3+} $(3d^6)$ and Mn^{3+} $(3d^4)$ have the same number of unpaired spins. If this were the only criterion, then both combinations should give zero magnetization, but in fact the oxide is ferrimagnetic with a substantial spontaneous magnetization.

The spinel ferrites, MFe_2O_4 where M is a divalent ion, have already been described: most of these have a more or less inverse population of the lattice sites represented by $(Fe^{3+})_A(M^{2+}Fe^{3+})_BO_4$, with the ferric ions equally distributed between octahedral and tetrahedral sites, and the divalent ions on octahedral sites. Cobalt ferrite, like magnetite, has been shown by neutron diffraction to be completely inverse [25]. Thus, the moments of the ferric ions cancel out, and the magnetization is contributed by the moments of the divalent ion. In cobalt ferrite the moment of $3.94\mu_B$ per formula unit (compared with $3\mu_B$ for the spin-only moment) cannot be explained by partial inversion. As in the other oxides the ionic moments of the Co^{2+} ions are anomalously high compared with the spin-only formulae.

It has been seen that when the angular momentum is completely quenched, the ionic magnetic moment is $2[S(S+1)]^{1/2}\mu_B$ (component $M_s\mu_B$ in ordered crystals) whereas for the free ion the moment becomes $g[J(J+1)]^{1/2}\mu_B$. When the angular momentum is not quenched, but the spin-orbit coupling is overcome by the crystal fields, the contributions from the orbital and spin angular momenta add vectorially so that $\mu^2 = \mu_L^2 + \mu_S^2$, that is

$$\frac{\mu^2}{\mu_B^2} = L(L+1) + 2^2 S(S+1) \ , \tag{4.57}$$

whence

$$p_e = [L(L+1) + 4S(S+1)]^{1/2} \ . \tag{4.58}$$

Table 4.4 compares the three possibilities with values of p_e, derived from the paramagnetic susceptibilities, for several 3d ions in crystals. As previously noted, the expression involving J is never good, except coincidentally for S-state ions. The spin-only formula is the one most generally applicable; however, cobalt is entirely distinctive in that the appropriate value of p_e is $[L(L+1)+4S(S+1)]^{1/2}$. A similar conclusion is drawn from the high (though variable) values of the components of the moments in ordered crystals: in other words cobalt is distinctive in that the orbital angular momentum is maintained, while it is substantially quenched in the rest of the 3d series. In view of the association of anisotropy with an interaction between the lattice and the orbital wavefunction, it is not surprising to find (as shown in the next section) that Co^{2+} is also the most anisotropic ion.

Table 4.4

Ion	Ground term	$[L(L+1)+4S(S+1)]^{1/2}$	$g[J(J+1)]^{1/2}$	$2[S(S+1)]^{1/2}$	Experimental
Ti^{3+}	$^2D_{3/2}$	3·00	1·55	1·73	1·7
Ti^{2+}	$^3F_2$	4·47	1·63	2·83	2·8
V^{2+}	$^4F_{3/2}$	5·20	0·70	3·87	3·8
Cr^{2+}	$^5D_0$	5·48	0·00	4·90	5·8–5·9
Mn^{2+}	$^6S_{5/2}$	5·92	5·92	5·92	5·8–5·9
Fe^{2+}	$^5D_4$	5·48	6·71	4·90	5·0
Co^{2+}	$^4F_{9/2}$	5·20	6·63	3·87	4·5–5·2
Ni^{2+}	$^3F_4$	4·47	5·59	2·83	3·2–3·5
Cu^{2+}	$^2D_{5/2}$	3·00	3·55	1·73	1·8

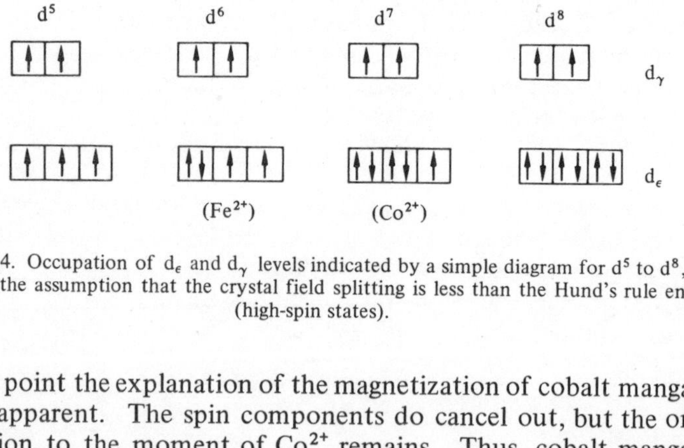

Figure 4.24. Occupation of d_ϵ and d_γ levels indicated by a simple diagram for d^5 to d^8, and based on the assumption that the crystal field splitting is less than the Hund's rule energy (high-spin states).

At this point the explanation of the magnetization of cobalt manganate becomes apparent. The spin components do cancel out, but the orbital contribution to the moment of Co^{2+} remains. Thus, cobalt manganate could be called an orbital ferrimagnetic.

The Bohr theory gives a clear mechanistic impression of angular momentum. A reasonable physical picture can also be obtained in terms of the orbitals representing the wavefunctions. In purely octahedral fields some of the orbitals are still energetically equivalent, and these may be interchangeable by rotation about a suitable axis. Thus, 90°

rotations transform the set d_{xy}, d_{xz} and d_{yz} into each other. Also, a $45°$ rotation of the d_{xy} orbital about the axis OZ produces the $d_{x^2-y^2}$ orbital, but these do not have the same energy in the crystal field and so such rotations of the charge cloud are suppressed. Hence, one gains a qualitative explanation of the partial quenching of the orbital angular momentum.

Now Co^{2+} (d^7) can be compared with its neighbours d^5, d^6 and d^9. Figure 4.24 shows the occupation of the energy levels by a simple diagram. For d^5 there can be no interchange of electrons even within the d_ϵ levels, since this would involve the superposition of orbitals with the same values of m and m_s, contravening the exclusion principle. The same applies to d^8, and clearly there can be no intercharge within the d_γ levels for any of the cases. Rotations, giving orbital angular momentum, can occur only for d^6 and d^7.

4.4.2 Anisotropy associated with cobalt ions

Van Vleck [26] noted that the anisotropy of the paramagnetic susceptibility (that is the ratio of the values measured along different crystal axes) varied considerably throughout the 3d series and was highest for cobalt, as shown in Table 4.5. The anisotropy of tetra-coordinated Co^{2+} is much smaller (5% for Cs_2CoCl_4) while the deviation from the spin-only formula for the ionic moment is also much smaller for this latter coordination. Thus, these results correspond to the general principle given above. Van Vleck distinguished between the effects in d^6 and d^7, Fe^{2+} and Co^{2+}, which would appear from the preceding section to behave similarly in an octahedral field, by considering the effect of adding a uniaxial component to the field.

Table 4.5†.

Iron	State		Anisotropy %
Cr^{3+} (d^3)	^4F	$Cr(NH_4)_3(C_2O_4)_3,3H_2O$	0·25
Mn^{2+} (d^5)	^6S	$Mn(NH_4)_2(SO_4)_2.6H_2O$	0·10
Fe^{3+} (d^5)	^6S	$FeK_3(C_2O_4)_3.3H_2O$	0·20
Fe^{2+} (d^6)	^5D	$FeK_2(SO_4)_2.6H_2O$	16·0
Co^{2+} (d^7)	^4F	$Co(NH_4)_2(SO_4)_2.6H_2O$	30·0
Ni^{2+} (d^8)	^3F	$Ni(NH_4)_2(SO_4)_2.6H_2O$	1·50
Cu^{2+} (d^9)	^2D	$Cu(NH_4)_2(SO_4)_2.6H_2O$	20·0

† From Van Vleck [26].

Figure 4.25 shows the additional splitting of the octahedral triplet by a uniaxial field, which might correspond to a slight distortion of a cubic crystal structure, rendering it tetragonal. For Co^{2+} the triplet is split into a doublet and a singlet, with the doublet lowest. For a D-state ion such as Fe^{2+} (d^6: ^5D) the lowest level is a singlet. If we assume population of the lowest level, then Co^{2+} can maintain its orbital momentum, whereas for Fe^{2+} the orbital momentum must be largely quenched.

A very large anisotropy is also associated with cobalt in the spinel ferrites, the effect of cobalt on the anisotropy of magnetite being particularly interesting. Manganese ferrite, nickel ferrite, and magnetite all have negative magnetocrystalline anisotropy constants of the same order of magnitude at room temperature, while cobalt ferrite has a positive and much larger K_1. Replacement of the divalent ions by cobalt may initially reduce the anisotropy and then cause it to change sign. In cobalt substituted magnetite, K_1 becomes positive at room temperature for exceptionally low concentrations of cobalt, and further substitution rapidly increases this positive anisotropy. Moreover, up to $x = 0 \cdot 15$ in the formula $Co_x Fe_{3-x} O_4$, the increase in K_1 is linear and suggests that the effect can be interpreted simply in terms of the anisotropy of the individual cobaltous ions, to which the other ions are, of course, coupled by the exchange interactions.

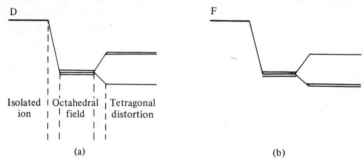

Figure 4.25. The way in which the lowest degenerate level left by an octahedral crystal field is further split by a tetragonal distortion, for ions such as Fe^{2+} (D-state) and Co^{2+} (F-state) (Van Vleck [26]).

The anisotropy constants, K_1, of some single ferrites and of cobalt-substituted ferrites are shown in Table 4.6 [27, 28].

Theoretical treatments of the effects of cobalt in magnetite have been given by Slonczewski [29] and reviewed by Kanamori [5]. In a distorted octahedral field the cobaltous ions, unlike Fe^{2+} for example, can still have an angular momentum with components $l_z = \pm\hbar$. The axis with which OZ is to be identified is found by examining the crystal symmetry. As shown by the part of the lattice drawn out in Figure 4.26, the immediately neighbouring anions give an octahedral environment, but the nearest cations superimpose a trigonal (3-fold rotational) axis with a $\langle 111 \rangle$ orientation. The angular momentum must have the same symmetry, that is to say the axis OZ is to be identified with the cube diagonals, and the components of the angular momentum

Table 4.6. The effect of cobalt on the anisotropy constant, K_1, of manganese ferrites[†] and of other ferrites[‡].

Composition	$10^{-3}K_1$ erg cm^{-3}	
	at 290°K	at 200°K
$MnFe_2O_4$	$-40 \cdot 7$	$-105 \cdot 3$
$Mn_{0 \cdot 91}Fe_{2 \cdot 09}O_4$	$-16 \cdot 8$	$-48 \cdot 5$
$Mn_{0 \cdot 99}Co_{0 \cdot 01}Fe_2O_4$	$-31 \cdot 2$	$-46 \cdot 8$
$Mn_{0 \cdot 96}Co_{0 \cdot 02}Fe_2O_4$	$-22 \cdot 2$	$+10 \cdot 0$
$Mn_{0 \cdot 96}Co_{0 \cdot 04}Fe_2O_4$	$-7 \cdot 0$	$+112 \cdot 2$
$Mn_{0 \cdot 94}Co_{0 \cdot 04}Fe_2O_4$	$+6 \cdot 3$	$+210 \cdot 0$
$Mn_{0 \cdot 92}Co_{0 \cdot 06}Fe_2O_4$	$+29 \cdot 1$	$-$
$Mn_{0 \cdot 90}Co_{0 \cdot 10}Fe_2O_4$	$+37 \cdot 3$	$+407 \cdot 5$
$Mn_{0 \cdot 73}Co_{0 \cdot 25}Fe_2O_4$	$+171 \cdot 0$	$-$
	20°C	-196°C
Fe_3O_4	-12×10^3	$-$
$Co_{0 \cdot 8}Fe_{2 \cdot 2}O_4$	$+3 \times 10^6$	$+4 \cdot 4 \times 10^6$
$Co_{1 \cdot 1}Fe_{1 \cdot 9}O_4$	$+1 \cdot 8 \times 10^6$	$-$
$Co_{1 \cdot 1}Fe_{2 \cdot 2}O_4$	$+3 \cdot 8 \times 10^6$	$+17 \cdot 5 \times 10^6$
$Co_{0 \cdot 3}Mn_{0 \cdot 4}Fe_2O_4$	$+1 \cdot 1 \times 10^6$	$-$
$NiFe_2O_4$	-51×10^3	$+20 \times 10^3$
$Co_{0 \cdot 002}Ni_{0 \cdot 7}Fe_{2 \cdot 2}O_4$	-18×10^3	-125×10^3
$Co_{0 \cdot 004}Ni_{0 \cdot 7}Fe_{2 \cdot 2}O_4$	-10×10^3	-196×10^3

† Pearson [27].   ‡ Bozorth, Tilden, and Williams [28].

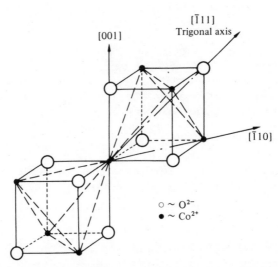

Figure 4.26. Taking the nearest neighbouring cations into account it is seen that an ion, which is octahedrally coordinated with the neighbouring anions, is also subjected to a crystal field with trigonal symmetry; the trigonal axis is parallel to $\langle 111 \rangle$ (Slonczewski [29]).

thus have this orientation. It would be tempting to conclude that the spin angular momentum, and magnetic moment, were similarly oriented, but this would not, after all, give the correct result since the positive anisotropy indicates that the spins lie parallel to $\langle 100 \rangle$. It is necessary to consider jointly the effects of the spin–orbit coupling and of the exchange field acting effectively on the spins to give an energy term $2\mu_B \mathbf{S} \cdot \mathbf{H}_e \propto 2\mu_B I_s H_e$ since I_s is the component of s parallel to $\mathbf{H}_e$. The energies corresponding to the alignment of the spins parallel either to a $\langle 100 \rangle$ direction or to a $\langle 111 \rangle$ direction can then be calculated, and it is found that the energy is least when the spins are aligned with $\langle 100 \rangle$ directions. Since the spin and orbital momenta are differently aligned, there should be a certain anisotropy of the spontaneous magnetization.

The above account is obviously simplified, and refers only to the effects of individual ions. It has been indicated that in fact this is adequate for the cobaltous ion, but anisotropy does arise from ions with non-degenerate ground states such as Fe^{2+} and Ni^{2+}. In these cases it is necessary to take account of the interactions between ions, and not just of the interaction between one ion and the lattice. Van Vleck [26] first considered the purely magnetostatic interaction between dipoles. The energy of interaction of dipoles μ_1 and μ_2 is

$$W = \frac{1}{r_{12}^3}\left[\mu_1 \cdot \mu_2 - \frac{3}{r_{12}^2}(\mu_1 \cdot \mathbf{r}_{12})(\mu_2 \cdot \mathbf{r}_{12})\right] , \qquad (4.59)$$

or, for identical dipoles, simply

$$W = \frac{\mu^2}{r^3}(1 - 3\cos^2\theta_{12}). \qquad (4.60)$$

For a cubic array of dipoles the mean value of $\cos^2\theta_{12}$ is $\frac{1}{3}$, and so $W = 0$ and there is no directional dependence. In addition, Van Vleck showed that the magnetostatic effects were generally much too small, although they can make the major contribution to the anisotropy of a few substances at very low temperatures, an example being the antiferromagnetic MnF_2.

A full analysis of the exchange interactions between pairs of cations, however, shows that this is in itself anisotropic. If we include the effects of spin–orbit interaction, the exchange Hamiltonian H_0, becomes $\hat{H}_0 + (\text{const.})\hat{H}_1 + (\text{const.})^2\hat{H}_2$; the first-order perturbation is

$$J\left(\frac{\lambda}{\Delta E}\right)^2\left[\mathbf{S}_i \cdot \mathbf{S}_j - 3\frac{(\mathbf{S}_i \cdot \mathbf{r}_{ij})(\mathbf{S}_j \cdot \mathbf{r}_{ij})}{r_{ij}^2}\right] , \qquad (4.61)$$

and the second order perturbation is

$$J\left(\frac{\lambda}{\Delta E}\right)^4 (\mathbf{S}_i \cdot \mathbf{r}_{ij})^2(\mathbf{S}_j \cdot \mathbf{r}_{ij})^2 . \qquad (4.62)$$

Thus, although the unperturbed Hamiltonian gives an isotropic exchange energy, the spin-orbit perturbations do give rise to energies which depend on the orientation of the spins with respect to the directions of the line joining them. Because of their form (compare Equations 4.59 and 4.61) these terms are said to represent pseudo-dipolar and pseudo-quadrupolar contributions to the anisotropy, but the interaction is electrostatic rather than magnetostatic in origin. These theories were put forward originally for metals.

In the above, ΔE is the difference in energy between the ground state and the first excited state, and λ the spin–orbit coupling constant. The theory shows that, for the quadrupolar term

$$K_1 \approx J\left(\frac{\lambda}{\Delta E}\right)^4 . \tag{4.63}$$

The difference between the value of g observed and that for spin, $(g-2)$, is also of the order of $\lambda/\Delta E$ and so Equation (4.62) can be written

$$K_1 \approx J(g-2)^4 . \tag{4.64}$$

For dipolar interactions

$$K_1 \approx J(g-2)^2 , \tag{4.65}$$

where J is the exchange integral corresponding to the super exchange interactions. This should apply, for example, to Ni^{2+} ions in oxides. However, it has been indicated that the major contribution to the anisotropy of nickel and manganese ferrites does not arise from the dipolar or quadrupolar interactions, but from the single-ion anisotropy of Fe^{3+} ions [32].

References

1. J. S. Griffith, *"The Theory of Transition Metal Ions"* (Cambridge University Press, Cambridge), 1961.
2. P. W. Anderson, *Phys. Rev.,* **79**, 350 (1950); **86**, 694 (1952); **115**, 2 (1959); *Solid State Phys.,* **14**, 99 (1963).
3. C. Zener, *Phys. Rev.,* **82**, 403 (1951).
4. J. B. Goodenough, *Phys. Rev.,* **100**, 564 (1955); **117**, 1442 (1960). J. B. Goodenough and A. L. Loeb, *Phys. Rev.,* **98**, 391 (1955).
5. J. Kanaromi, in *"Magnetism",* Volume 1, Eds. G. T. Rado and H. Suhl (Academic Press, New York), 1963.
6. G. Blasse, *Philips Res. Rept.,* **18**, 383 (1963); *Philips Res. Rept. Suppl.* 3 (1964); *Phys. Chem. Solids,* **27**, 383 (1966).
7. J. Smit and H. P. J. Wijn, *Advan. Electron. Electron Phys.,* **6**, 70 (1954).
8. H. A. Mook and C. G. Shull, *J. Appl. Phys.,* **37**, 1034 (1966).
9. W. Pauli, *Z. Physik,* **41**, 81 (1927).
10. R. T. Schumacher and C. P. Slichter, *Phys. Rev.,* **101**, 58 (1956).
11. R. T. Schumacher, T. R. Caver and C. P. Slichter, *Phys. Rev.,* **95**, 1089 (1954).
12. D. Pines, *Phys. Rev.,* **95**, 1090 (1954).
13. J. M. Ziman, *"Electrons in Metals"* (Taylor and Francis, London), 1963; *"Principles of the Theory of Solids"* (Cambridge University Press, Cambridge), 1964.

14. S. Taniguchi, R. S. Tebble and D. E. G. Williams, *Proc. Roy. Soc.,* **A265,** 502 (1962).
15. E. C. Stoner, *Proc. Roy. Soc.,* **A154,** 656 (1936); **A165,** 372 (1938).
16. E. P. Wohlfarth, *Proc. Roy. Soc.,* **A195,** 434 (1948); *Phil. Mag.,* **40,** 1095 (1949); *Phil. Mag.,* **45,** 647 (1954).
17. N. F. Mott, *Advan. Phys.,* **13,** 325 (1964).
18. M. A. Ruderman and C. Kittel, *Phys. Rev.,* **96,** 99 (1954).
19. S. V. Vonsovskii, *J. Phys. U.S.S.R.,* **10,** 468 (1946).
20. C. Zener, *Phys. Rev.,* **81,** 440 (1951); **83,** 299 (1951); C. Zener and R. R. Heikes, *Rev. Mod. Phys.,* **25,** 191 (1953).
21. R. Stuart and W. Marshall, *Phys. Rev.,* **120,** 353 (1960).
22. R. E. Watson, *Phys. Rev.,* **118,** 1036 (1960).
23. W. L. Roth, *Gen. Elec. Res. Lab. Rept.,* **63** (1963).
24. P. F. Bongers, Thesis, Leyden, 1957.
25. C. G. Shull, *J. Phys. Radium,* **20,** 169 (1959).
26. J. H. Van Vleck, *Discussions Faraday Soc.,* **26,** 96 (1958).
27. R. F. Pearson, *J. Appl. Phys.,* **31,** 160S (1960).
28. R. M. Bozorth, E. F. Tilden and A. J. Williams, *Phys. Rev.,* **99,** 1788 (1955).
29. J. C. Slonczewski, *J. Appl. Phys.,* **32,** 253S (1961).
30. H. Danan, A. Herr and A. J. P. Meyer, *J. Appl. Phys.,* **39,** 669 (1968).

5

Effects of Crystal Size and Shape
Rotational Processes

This chapter is concerned with the response to applied fields of the magnetization of ferromagnetic or ferrimagnetic specimens which are not subdivided into domains. Thus the magnetization is basically uniform in both magnitude and direction, although important deviations from this rule occur. The magnetization can only be changed by rotations of I_s, and not by wall motion; if the magnetization vectors rotate as a whole, with all the spins remaining parallel, the effect is known as *coherent rotation*. In the case of *incoherent rotation* (as in wall motion) the spins do not remain parallel but vary in direction from place to place in the crystals.

Rotation is the most basic and best understood type of magnetization process. It is relatively insensitive to the state of crystalline perfection of the specimens, as compared with domain wall motion, although it is strongly influenced by particle shape.

In the first part of this chapter it is assumed that thermal equilibrium is always attained. Dynamic effects and the approach to equilibrium are discussed in the latter part. Furthermore, it is initially assumed that rotation is the only relevant mode of magnetization change, and the conditions under which this is to be expected, rather than domain wall nucleation and motion, are examined in section 5.8.

5.1. COHERENT ROTATION AGAINST UNIAXIAL ANISOTROPY

It will be assumed first that uniaxial anisotropy can be represented by the energy approximation involving a single constant (per unit volume)

$$W_K = -K\sin^2\theta , \tag{5.1}$$

and the corresponding torque

$$T_K = -\frac{\partial W_K}{\partial \theta} = 2K\sin\theta\cos\theta ; \tag{5.2}$$

or it may be represented by an effective anisotropy field $H_K = 2K/I_s$ for small values of θ, the angle between I_s and the easy direction. Some of the features of coherent rotation are most clearly demonstrated by finding the direction of magnetization for which the 'restoring' anisotropy torque is equal to the 'driving' torque due to an applied field, that is to

$$T_H = \mathbf{H} \times \mathbf{I}_s = HI_s \sin\phi .$$

(see Figure 5.1a).

Initially the anisotropy can be assumed to be due to crystal anisotropy in a uniaxial crystal of exactly spherical shape, in order to make the demagnetizing energy independent of the direction of the magnetization. The simplest analysis can be made when the field is perpendicular to the easy axis as in Figure 5.1a:

$$T_H = HI_s \sin\phi = HI_s \cos\theta \tag{5.3}$$

and the induced magnetization, which is the component of I_s parallel to the field, is

$$I = I_s \cos\phi = I_s \sin\theta . \tag{5.4}$$

For any value of H the equilibrium orientation of I_s is given by Equations (5.2) and (5.3) ($T_K = T_H$) as

$$\theta = \sin^{-1}\frac{HI_s}{2K} .$$

The corresponding magnetization is

$$I = I_s \sin\theta = \frac{HI_s^2}{2K} , \tag{5.5}$$

and the susceptibility is constant,

$$\chi = \frac{I_s^2}{2K} , \tag{5.6}$$

up to saturation in a field H_s given by putting $I = I_s$ in Equation (5.4):

$$H_s = 2K/I_s \quad (= H_K) . \tag{5.7}$$

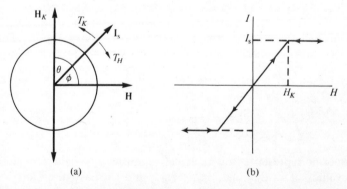

(a)   (b)

Figure 5.1. (a) A uniformly magnetized spherical crystal with a single easy axis represented by the direction of the anisotropy field, $\mathbf{H}_K$. (b) The reversible magnetization curve for the model shown in (a).

Hence, the complete behaviour of this simple system is readily calculable and can be represented by the magnetization 'loop' in Figure 5.1b. This is completely reversible; there is no hysteresis and H_c and I_r are zero.

A contrasting situation, with **H** antiparallel to $\mathbf{H}_K$, can be approached in a different way. Somewhat remarkably, the field exerts no torque whatsoever in the case of ideal alignment and structural perfection. However, when $H > H_K$ $(= 2K/I_s)$ the net field is 'downwards' and only the slightest perturbation is needed to initiate reversal. By plotting the torques given by Equations (5.2) and (5.3), and with $\theta = \phi$ for this case, it can be seen that with $H > H_K$ the initial equilibrium is meta-stable and after a small displacement the net torque is always towards **H**. Thus, the magnetization loop is as shown in Figure 5.2b; I_r equals I_s and the magnetization reverses completely in a single discontinuity at $H = H_K$, which is the coercivity or, in this case, switching field.

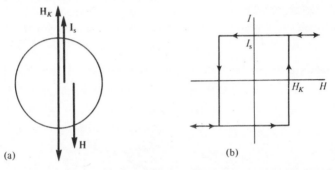

(a) (b)

Figure 5.2. (a) A spherical uniaxial crystal with a field applied antiparallel to the initial direction of magnetization. (b) The magnetization loop for (a), with complete reversal of the magnetization when the value of the applied field becomes equal to H_K.

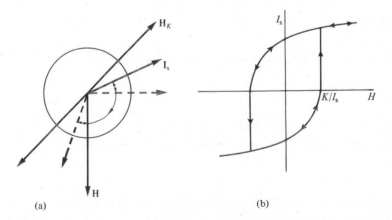

(a) (b)

Figure 5.3. For the situation shown in (a), the magnetization loop (b) is partly reversible and partly irreversible.

For certain orientations, such as that shown in Figure 5.3, the loops can easily be seen to be partly reversible and partly discontinuous, giving a net hysteresis. With the sign convention for the fields as indicated, the general form of the loop shown in Figure 5.3b can be derived by inspection, and the precise loop can readily be computed numerically. Clearly $I_r = I_s \cos 45° = I_s/\sqrt{2}$. Beyond $\theta = 45°$, T_H increases with θ while T_K decreases, and the discontinuity occurs at H_c which is given by

$$H_c I_s (\sin 90°) = 2K \sin 45° \cos 45° = K$$

$$H_c = \frac{K}{I_s} . \tag{5.8}$$

5.2 FURTHER DEVELOPMENT: SHAPE ANISOTROPY

Although all the most important features of switching by coherent rotation can be demonstrated quite simply as above, a full treatment is quite lengthy, particularly when assemblies of randomly oriented crystals or crystallites are included. The original calculations were carried out by Stoner and Wohlfarth [1], who minimized the sum of the energies W_K and W_H to determine the equilibrium orientations of the magnetization, and set the second derivative to zero to determine the conditions for stability.

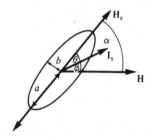

Figure 5.4. An ellipsoidal crystal, assumed to have neither crystal nor stress anisotropy, but with unequal principal demagnetizing factors (N_a applying to the major axis and N_b to the minor axis: $N_b > N_a$) which give an effective shape anisotropy.

If we take the example shown in Figure 5.4 and assume that the material has no crystal anisotropy, no magnetostriction, and thus no stress anisotropy, then only the demagnetizing energy, W_D, need be considered. With the angle θ between the magnetization direction and the major axis this is given by

$$W_D = \tfrac{1}{2} I_s^2 (N_a \cos^2\theta + N_b \sin^2\theta)$$
$$= \tfrac{1}{4} I_s^2 (N_a + N_b) - \tfrac{1}{4} I_s^2 (N_b - N_a) \cos 2\theta , \tag{5.9}$$

which is minimum when θ equals 0 or π, and maximum when θ equals $\pi/2$ (so long as $N_a < N_b$ as shown) and thus indicates a shape anisotropy.

After adding the term $(-HI_s \cos\phi)$ and noting that $\theta = \alpha - \phi$, the equilibrium relation is obtained by differentiating the total energy as

$$\frac{\partial W}{\partial \phi} = \tfrac{1}{2}I_s^2(N_b - N_a)\sin 2\theta + HI_s \sin\phi$$
$$= 0. \qquad (5.10)$$

This treatment is equivalent to equating the torques due to the applied field and to the shape anisotropy. For small angles $\sin 2\theta \doteqdot 2\sin\theta$ and thus we may define a shape anisotropy field, H_{sh}, by

$$[(N_b - N_a)I_s]I_s \sin\theta = H_{sh}I_s \sin\theta ,$$

that is

$$H_{sh} = (N_b - N_a)I_s . \qquad (5.11)$$

For the sake of universality in displaying the results it is useful to introduce a reduced field, h, defined by $h = H/H_{sh} = H/(N_b - N_a)I_s$. If we replace θ by $(\phi - \alpha)$ Equation (5.10) becomes

$$\tfrac{1}{2}\sin 2(\phi - \alpha) + h \sin\phi = 0, \qquad (5.12)$$

which can be solved numerically to give ϕ in terms of h, and hence the magnetization curves shown in reduced terms in Figure 5.5, since $I = I_s \cos\phi$. The results shown are for a series of values of α, and three of these are identical to those derived by the approximate method using a crystalline anisotropy field, if h is alternatively identified with $H/H_K = HI_s/2K$. It is easy to see that the treatment of the two cases is

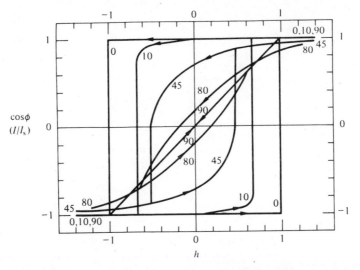

Figure 5.5. Magnetization curves for various values of the angle between the major axis and field direction (as in Figure 5.4) shown in degrees by the numbers on the curves. The reduced field $h = H/(N_b - N_a)I_s$, and the same curves apply to spherical crystals with uniaxial crystal anisotropy if $h = H/H_k = HI_s/2K_1$.

identical so long as the anisotropy energy is represented by $K\sin^2\theta$ only, that is the higher order terms are neglected. Again, if only magneto-elastic energy is taken into account ($W = \frac{3}{2}\lambda\sigma\sin^2\theta$) the same results apply with

$$h = \frac{HI_s}{3\lambda\sigma} \,. \qquad (5.13)$$

It is clear from Figure 5.5 that the maximum values of the coercivity are obtained when the field is applied parallel to the easy axis, and are equal to the anisotropy fields. Figure 5.6 shows a computed magnetization curve for a randomly oriented assembly of particles. The remanence is $0\cdot5I_s$ and the coercivity is

$$H_c = 0\cdot479(N_b - N_a)I_s \,. \qquad (5.14)$$

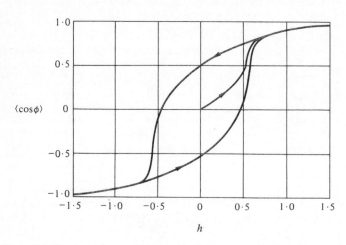

Figure 5.6. Magnetization curve for an assembly of ellipsoids oriented at random. $\langle\cos\phi\rangle$ is proportional to the magnetization [1].

5.3 SUSCEPTIBILITIES CORRESPONDING TO ROTATIONAL PROCESSES

It has already been shown that when a field is applied at right angles to a single easy axis, the susceptibility is constant up to saturation: $\chi = I_s^2/2K$. This result was obtained on the assumption of a single anisotropy constant. When the anisotropy is described by two constants, the susceptibility is no longer constant. It is easy to find the equilibrium values of θ by differentiating ($W_K + W_H$) after using

$$K_1\sin^2\theta + K_2\sin^4\theta = \tfrac{1}{2}K_1 + \tfrac{3}{8}K_2 - \tfrac{1}{2}(K_1 + K_2)\cos2\theta + \tfrac{1}{8}K_2\cos4\theta \,. \qquad (5.15)$$

Since $I = I_s \sin\theta$ again, $\sin\theta$ can be eliminated to give

$$HI_s = 2\frac{I}{I_s}\left[K_1 + 2K_2\left(\frac{I}{I_s}\right)^2\right] \qquad (5.16)$$

The magnetization curve and the susceptibility at any point can be computed from this expression, and good agreement is obtained with magnetization curves measured normal to the easy axis of cobalt, for example.

For initial susceptibilities only, I/I_s is small and thus $(I/I_s)^2$ can be neglected in Equation (5.16) above, giving

$$HI_s = \frac{2K_1 I}{I_s} \ ;$$

whence

$$\chi_0 = \frac{I}{H} = \frac{I_s^2}{2K}$$

once more. (It is assumed that K_2 is never much greater than K_1; the approximation is clearly equivalent to neglecting $\sin^4\theta$.) The most useful expressions generally for initial susceptibilities are those for randomly oriented assemblies of crystals or crystallites, since these represent the most common practical situation encountered. Using the single constant approximation, for a field oriented at any angle α to an easy direction one obtains the (single particle) initial susceptibility as

$$\chi_0 = \frac{I_s^2 \sin^2\alpha}{2K_1} \qquad (5.17)$$

and the same expression applies to a crystal with cubic anisotropy and $K_1 > 0$, such as iron. For a cubic crystal with $K_1 < 0$ (nickel and most ferrites) the anisotropy energy for small deflections is $W_K = -\frac{2}{3}K_1\theta^2$ or $\frac{2}{3}|K_1|\theta^2$ and this gives

$$\chi_0 = \frac{3I_s^2 \sin^2\alpha}{4|K_1|} \ . \qquad (5.18)$$

For the assembly, χ_0 is obtained by use of the mean value of $\langle\sin^2\alpha\rangle = \frac{2}{3}$ as

$$\chi_0 = \frac{I_s^2}{3K_1} \ \text{(uniaxial, } K_1 > 0) , \qquad (5.19)$$

$$\chi_0 = \frac{I_s^2}{2K_1} \ (K_1 < 0) \ . \qquad (5.20)$$

One important implication arising from these formulae is purely negative. Inserting values for iron for example, gives $\chi_0 \doteqdot 30$, compared with measured values ranging up to 10^4. Many similar observations lead to the conclusion that rotational processes are the exception for bulk materials; the susceptibilities calculated for coherent rotation can generally be regarded as lower limits, and are found in practice to apply only to small 'single-domain' particles or crystallites. There are materials, such as nickel–iron alloys or manganese–zinc ferrites, that can be produced with anisotropies which pass from positive to negative values as the composition or heat treatment is adjusted, and in principle it is possible that $K_1 \to 0$ and $\chi_0 \to \infty$ for rotation. Even in these cases, however, it seems that there is always sufficient residual anisotropy, of some form, to favour alternative magnetization processes (see Chapter 6).

At very high frequencies, however, domain wall motion is heavily damped, and rotation does make the main contribution to magnetization changes.

5.4 REMANENCE

The ideal remanence of a single uniaxial crystal, or oriented assembly of crystals, is zero after magnetization in fields normal to the easy axes. After saturation in fields parallel to the easy axes the remanence measured in this direction is I_s, or unity in reduced units of I_r/I_s.

For a randomly oriented assembly of uniaxial particles, or polycrystal, it may be assumed that on the removal of a saturating field, the magnetization relaxes back from the field direction, through an angle α, to the nearest easy direction to this field direction. The remanence is then $I_s\langle\cos\alpha\rangle$, where α varies between 0 and $\pi/2$ and the average must be taken three-dimensionally. The result is just $0\cdot5I_s$, as is indicated by the loop in Figure 5.6.

It is to be expected that the remanence should be higher for cubic materials, since there is then a higher probability of an easy direction lying close to the field direction. The averages were worked out by Gans [2], and gave

Type of anisotropy	uniaxial	cubic $K_1 > 0$	cubic $K_1 < 0$
Remanence, I_r/I_s	$0\cdot5$	$0\cdot832$	$0\cdot866$

5.5 THE EFFECT OF PARTICLE SIZE ON REVERSAL MECHANISMS AND COERCIVITY

So far, no reference has been made to particle size, and it is now necessary to consider in what circumstances coherent reversal should occur and the properties be those forecast for this behaviour as above. In fact, these circumstances are quite restricted.

5.5.1 Superparamagnetism

The first assumption to be considered is that, in the absence of a field, the magnetization has a stable orientation due to the anisotropy. For a particle of volume v, this energy can be generalized as $E_K = \frac{1}{2}Cv\sin^2\theta$ where $C = 2K_1$, $(N_b - N_a)I_s^2$ or $3\lambda\sigma$, and it has a maximum value of $\frac{1}{2}Cv$ as θ goes through $\pi/2$ for a uniaxial particle. In fact, the direction of magnetization must fluctuate continuously under the influences of thermal agitation, and if the particle volume and temperature are such that $\frac{1}{2}Cv \doteq kT$, there is a finite probability that the magnetization will be spontaneously reversed.

The theoretical treatment due to Néel [3] and Brown [4] introduces a relaxation time, τ, for this decay of the magnetization with time in a particle assembly after the removal of a saturating field, that is:

$$I(t) = I_s e^{-t/\tau}. \tag{5.21}$$

This expression can also be regarded as a representation of the decay of the remanence. In the usual way $1/\tau$ is the probability per second that the magnetization will be reversed by thermal agitation, and is thus related to the activation energy required ($\frac{1}{2}Cv$) by the Boltzmann relation

$$\frac{1}{\tau} = f \exp\left(-\frac{Cv}{2kT}\right). \tag{5.22}$$

This is a standard statistical treatment. The great problem is to identify the frequency factor f. It is not unreasonable to suppose that it results from precession, since the precessional frequency represents the number of times per second that the energy barrier is approached. Consequently, $f = (\gamma/2\pi)H$; $\gamma \doteq 10^7$, and H, by a rather difficult argument, is identified with the coercive field (that is the coercivity at absolute zero or by neglect of thermal fluctuations) which gives $f \doteq 10^9$.

It is now possible to make an arbitrary definition of a superparamagnetic system, as one in which the remanence decays at a readily apparent rate, say with τ equal to 10^2 seconds. If numerical values are inserted with T set at $300°$K and only crystal anisotropy is assumed to be effective (spherical particles), one obtains critical radii of about 100 Å and 50 Å for iron and cobalt respectively. These sizes have quite a definite meaning, for it is easy to show that the exponential function varies very rapidly as the index changes: thus, for radii a little smaller than the above the decay is very rapid, whereas a small fractional increase in the radii gives values of τ so great that they may be taken to represent complete stability.

Magnetic viscosity is an effect closely associated with the above. Suppose that a field a little below the coercive field H_0 is applied to an assembly of stable particles. In spite of the stability in the absence of a field, thermal activation may cause reversals if kT is comparable to the required activation energy, $(H - H_0)vI_s$. An expression very similar to Equation (5.22) is obtained and, if $1/\tau$ is again equated to the probability of reversal of the magnetization of identical particles in unit time, the expression is also proportional to the rate of change of magnetization, giving

$$\frac{\mathrm{d}I}{\mathrm{d}T} \propto \log H \ . \tag{5.23}$$

It will be seen that $\mathrm{d}I/\mathrm{d}t \propto (H-H_0)$ for eddy current or relaxation damping, and so Equation (5.23) is important in a diagnostic sense.

The most obvious test for super paramagnetism is the absence of remanence, or zero coercivity. A further test is to plot the magnetization measured at different fields and temperatures as a function of H/T: the plots should be linear and saturation magnetization values can be obtained by extrapolating for H/T tending to infinity. Figure 5.7 shows a reversible magnetization curve for nickel crystallites with mean diameter of 100 Å [5].

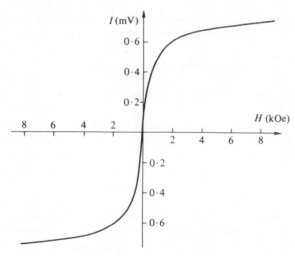

Figure 5.7. Reversible magnetization curve for super paramagnetic nickel: particle diameter is approximately 100 Å. Magnetization in arbitrary units.

5.5.2 Incoherent rotation

A second implicit assumption in the treatment of coherent rotation is that the exchange energy is of such overriding importance that the magnetization can be represented by a single vector. Although exchange forces are very powerful they are not infinitely so, and it can be seen from Figure 5.8 that the energy peak corresponding to coherent rotation against shape anisotropy can be reduced at the expense of some exchange energy.

In the most general type of calculation it is necessary to minimize the sum of the energy terms for the magnetization in the applied field, the demagnetizing or stray field energy, the crystal anisotropy energy *and the exchange energy* for a magnetization distribution which varies continuously in direction. This total free energy will have a form

$$E_T = \int \{ -\mathbf{I} \cdot \mathbf{H} - \tfrac{1}{2} \mathbf{I} \cdot \mathbf{H_d} + W_K + A[(\nabla \alpha_1)^2 + (\nabla \alpha_2)^2 + (\nabla \alpha_3)^2] \} \, dv \,, \quad (5.24)$$

and the minimization can only be carried out numerically, or in restricted conditions, and by approximate methods. Thus, it is only practicable to quote certain of the results, and reference should be made to the original papers [6] and the monographs by Brown [7].

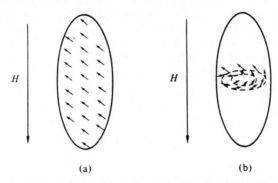

Figure 5.8. Magnetization reversal in an ellipsoid by coherent rotation (a) and by curling (b) for which the shape anisotropy is effectively reduced while some exchange energy arises.

It is first noted empirically that a new and important basis of size-dependence of magnetic behaviour (particularly coercivity) arises when the direction of magnetization is allowed to vary. The exchange energy depends on the rate of change of the direction of magnetization, and thus a non-uniform mode is more likely the larger the particles. It is found that incoherent rotation may significantly reduce the coercivity, although in the same situation the reversible response of the magnetization to small fields remains coherent rotation, and the initial susceptibilities are unaffected.

A second critical size can now be defined by that value of b for which curling is replaced by coherent rotation and $H_c \to H_s$. According to Equation (5.27) this happens when the last two terms are equal and

$$b_c = \left(\frac{2\pi k A}{N_a I_s^2} \right)^{1/2} . \quad (5.25)$$

The value of b_c ranges from $1 \cdot 04 (A/I_s^2)^{1/2}$ for the long cylinder to $1 \cdot 44 (A/I_s^2)^{1/2}$ for the sphere. For high values of I_s it can be very close to the critical size for superparamagnetic behaviour, depending on the other parameters involved, and in this case the expressions for super paramagnetism should be modified to account for the reduction in the energy barriers.

5.5.3 Fanning

If a chain of n spheres forms a linear aggregate, it can be approximated to an ellipsoid with an axial ratio of value n, and thus the aggregate can have a shape anisotropy even when each individual sphere is isotropic. If rotation is coherent for the whole aggregate, the maximum shape anisotropy barrier is involved and $H_c = (N_b - N_a)I_s$. However, there is an alternative mode known as fanning, illustrated by Figure 5.9, which appears to involve a lower energy barrier and thus to be most likely to give rise to the coercivity in practice: this can be seen by noting that the self energy of each sphere is independent of the direction of magnetization, while the interaction energy of two neighbouring spheres is much less if the magnetization vectors are antiparallel than if they are parallel. A simple analysis of this interaction problem [8] gave the coercivity for fields applied parallel to the chain length as

$$H_c = a\pi I_s , \tag{5.26}$$

with a ranging from $0 \cdot 5$ for two spheres to approaching $1 \cdot 2$ as $n \to \infty$. (Note that $H_c = 2\pi I_s$ for coherent rotation in an infinite cylinder.)

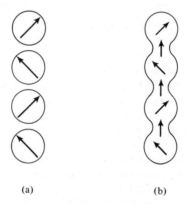

(a)   (b)

Figure 5.9. (a) The rotation of the magnetization in opposite senses in alternate elements of a chain of spheres, known as *fanning*. Fanning may also occur in irregularly shaped single particles as shown in (b) in which case some exchange energy must clearly be involved.

Fanning, as described above, does *not* give a size-dependent coercivity. However, a modified and more complex form of fanning may occur in continuous elongated particles which are simply irregular in shape or which are interconnected to form an irregular network (Figure 5.9b). In these cases some exchange energy is involved and some size-dependence is to be expected.

5.5.4 Curling

The curling mode of reversal is as illustrated by Figure 5.8. The basic feature is the reduction of the energy barrier, otherwise corresponding to the shape anisotropy, at the expense of the exchange energy introduced. The extensive review by Wohlfarth [9] shows that the coercivity of a prolate spheroid with negligible crystalline or strain anisotropy is

$$H_c \geqslant \frac{2\pi k A}{b^2 I_s} - N_a I_s \;. \tag{5.27}$$

A is the exchange constant, N_a the demagnetizing factor parallel to the major semi-axis and the applied field direction, and the size-dependence is indicated by b, the minor semi-axis. The constant k ranges from $1 \cdot 38$ for a sphere to $1 \cdot 08$ for a long cylinder ($a \gg b$). For the latter the equality sign applies (and $N_a = 0$):

$$H_c = \frac{6 \cdot 78 A}{b^2 I_s} \;. \tag{5.28}$$

The effect of crystal anisotropy, when it is collinear with the shape anisotropy, can be represented by

$$H_c \geqslant \frac{2K}{I_s} + \frac{2\pi k A}{b^2 I_s} - N_a I_s \;, \tag{5.29}$$

but this only applies so long as K is small and shape anisotropy is predominant.

5.5.5 Experimental evidence for incoherent rotation

Several types of experiment indicate the existence of incoherent rotation. One of the most important is the variation of the switching field or coercivity with particle size. Figure 5.10 shows some experimental results [10] from which it is inferred that the magnetization of the whiskers is reversed by a curling mechanism. For elongated single domain particles, however (see below), the independence of H_c over a considerable range of sizes indicates reversal by fanning.

The coercivity for coherent rotation is also independent of crystal size. Other experiments are needed to distinguish between this and fanning, although the magnitude of H_c may be sufficient. One way is to measure H_c whilst varying the angle between the applied field and the easy direction and to compare the results with the curves calculated for coherent rotation and for fanning: Figure 5.11 requires little comment.

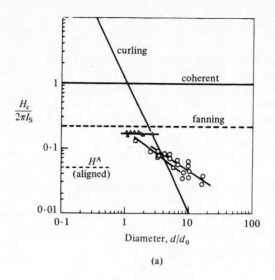

(a)

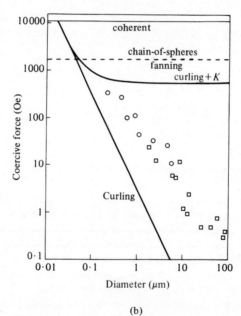

(b)

Figure 5.10. The variation of coercivity with particle diameter for:

(a) e.s.d. particles △, and iron wire ○ or elongated iron particles from the reduction of acicular iron oxides □. $d_0 = 2A^{1/2}/I_s = 120$ Å for both iron and iron cobalt. The crystal anisotropy field for aligned iron crystals is indicated as H^A (Luborsky [11]).

(b) Iron whiskers: □ measurements by Meiklejohn, quoted by authors. The line marked 'curling + K' is for the combined effects of shape and crystal anisotropy. It is indicated that the e.s.d. particles switch by a fanning mechanism; the others by curling (Luborsky and Morelock [10]).

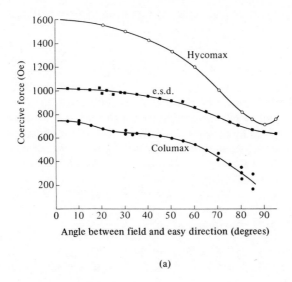

(a)

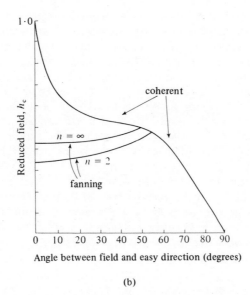

(b)

Figure 5.11. (a) The measured variation of coercivity with the angle between the field direction and the easy axis for three anisotropic materials. Calculated curves for coherent rotation (b) show a sharp drop in h_c at small angles, and it is indicated that fanning occurs in all the materials in (a).

5.6 SHAPE ANISOTROPY AND PERMANENT MAGNET MATERIALS

Many permanent magnet alloys, of the precipitate-hardened type, can be considered as assemblies of elongated single-domain particles, as first discussed extensively by Stoner and Wohlfarth [1] (the original paper should be regarded as essential reading for anyone interested in this particular topic). The so-called e.s.d. (elongated single domain) materials represent a very successful attempt to synthesize the type of structure found in the alloys [11]. This is done by producing masses of cobalt–iron particles, with the shape indicated by Figure 5.9b, using controlled electrodeposition into mercury; the particles are aligned by a magnetic field and the mercury replaced by a non-magnetic matrix which will form a rigid compact. Technical advantages include the ability to press magnets to the finally desired shape, without the necessity for machining.

The basis for the opening statement of this section is supplied by Figure 5.12 (facing p.212). This is an electron micrograph of a replica of a most ingenious type prepared by de Vos [12] from the surface of a single crystal of Ticonal XX, a typical high-quality precipitation alloy with the composition 35% Fe, 34·8% Co, 14·9% Ni, 7·5% Al, 2·4% Cu, 5·4% Ti. The precipitate particles are strongly magnetic and rich in iron, while the non-magnetic (or weakly magnetic) matrix contains most of the aluminium. After careful surface treatment a layer of aluminium oxide forms over the matrix, and this can be removed for electron microscopic examination to give a direct indication of both the morphology and composition of the two phases. The regularity of the array of close-packed magnetic particles shown in this figure is remarkable, and this appears to have a subtle connection with the inclusion of titanium in the composition. Other alloys of the same basic type have less regular microstructures, and there may be many more cross links between the particles. After first producing a single phase by heating at about 1100–1200°C, the finely divided structure is obtained by tempering the alloys at 500–600°C: the coercivity rises at first as the precipitate forms and then falls as the particles coarsen and, presumably, some form of incoherent rotation becomes increasingly effective. Coercivities of different alloys vary from 600 Oe to over 1000 Oe. However, a high coercivity is far from being the only requirement of a good permanent magnet.

When magnets can be used in a form giving rise to low demagnetizing fields (as for compass needles) the principal requirement is a high remanence and thus a high I_s, and a relatively low coercivity may be acceptable: hard steels, in which the limited coercivity may be associated with impediments to domain wall motion rather than effective fine particles, may be employed. More usually a permanent magnet will be required to maintain large fields either in a substantial air gap, or when it is in the form of a short bar or rod; in both cases it will then be exposed to considerable demagnetizing fields and a high coercivity will be required to maintain a high level of magnetization.

Assuming that the field in the gap of the ring magnet shown in Figure 5.13 is approximately uniform, the total magnetic energy stored in the gap is $(1/8\pi)H_g^2 V_g$. This energy can be related to the product of the induction and field in the magnetic material, that is to the 'energy product', BH, and so this becomes a general specification of the usefulness of the material. Since the flux must be continuous, we have

$$B_g A_g \, (= H_g A_g) = B_m A_m \ . \tag{5.30}$$

Also, since there are no macroscopic currents $\int \mathbf{H} \cdot \mathbf{dl} = 0$, and taking the path shown,

$$H_g l_g = -H_m l_m \ , \tag{5.31}$$

with the rather doubtful assumption that H_m is constant throughout. These two equations give

$$H_g^2 V_g = B_m H_m V_m \tag{5.32}$$

and the stored energy is greatest when the energy product for the material is greatest. (This is a standard treatment, but it does seem a little dubious in some respects.)

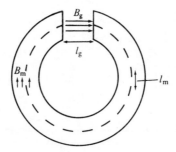

Figure 5.13. A model ring magnet used for the calculation of energy products. B_m and B_g are the induction in the magnetic material and in the gap respectively, and l_m and l_g the length of the path through the material and the gap, along which the integration is carried out. The sectional area of the material is A_m and that of the gap A_g.

The next assumption made in the technological specifications is that it is unnecessary to give B/H loops for a large range of different shapes and thus for different demagnetizing factors, but that it is possible to work simply with the B/H loop (or demagnetizing curve) as measured in a magnetic circuit with no demagnetizing factor. The effect of different demagnetizing fields is then taken to be equivalent to that of the reverse fields applied during the measurement: although the situations are far from being equivalent, this seems to be generally acceptable.

Consider an assembly of perfectly aligned long cylindrical particles with a low packing density p, and of such a size that reversals are coherent. The coercivity should then approach $2\pi I_s$, and the saturation induction will be $4\pi p I_s$. Assuming that no reversals occur until the coercivity is reached, the ideal demagnetizing curve will be as shown in Figure 5.14, approaching $B = 0$ at $H = -4\pi p I_s$; that is the sloping line is given by $B = 4\pi p I_s - H$ and

$$BH = 4\pi p I_s H - H^2. \tag{5.33}$$

For the maximum value

$$\frac{\partial}{\partial H}(BH) = 4\pi p I_s - 2H = 0 .$$

which occurs at

$$H = 2\pi p I_s ,$$

so that the plot of the energy product is roughly as shown. Thus, the maximum value of BH, $(BH)_{max}$, can be achieved even if $H_c < 2\pi I_s$.

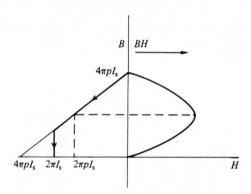

Figure 5.14. Ideal demagnetizing curve for non-interacting aligned particles with $H_c = 2\pi I_s$, packing fraction p, and perfectly square magnetization loops.

Suppose now that the density of packing, p, is increased. Obviously B_s rises in proportion to p, but H_c must fall, for in the limit $p \to 1$ the identity of the particles is lost. Unfortunately there is no analytical formula for the effect of p on H_c, but to optimize $(BH)_{max}$ it is clearly advisable to increase the packing until $H_c = 2\pi p I_s$.

Only a limited amount of control can be achieved in the alloys, but it is not surprising to find that different heat treatments are used to give either the maximum coercivity or the maximum $(BH)_{max}$.

Great technical effort has been devoted to the problem of the orientation of the precipitate particles. When tempering is carried out in the absence of a magnetic field the rods or platelets grow along the principal axes of the cubic crystal structure, and give an isotropic alloy with relatively low remanence and energy product (in principle $I_r = 0 \cdot 5I_s$). If a field is applied during cooling or tempering, growth along the $\langle 100 \rangle$ directions nearest to the field direction is encouraged, giving anisotropic 'equi-axed' alloys. For these the remanence should be that calculated for cubic crystal anisotropy (K_1 positive, $I_r = 0 \cdot 832I_s$) and the energy product is also improved. Even better results, as indicated by the demagnetizing curves in Figure 5.15, are obtained by producing columnar crystallites, with a common [100] axis. This can be done by casting the alloys into a mould with a chilled base, and a specimen with perfectly-oriented columnar crystallites should be equivalent to a single crystal [13].

An indication of the complexity of the technology is given by recent work on alloys containing titanium. Titanium raises the coercivity of the alloys, and since the effect on the saturation induction is negligible very high values of $(BH)_{max}$ are also obtained. However, it was at first accepted that columnar titanium alloys could not be produced, because of the 'grain refining' effect of titanium. Later developments then

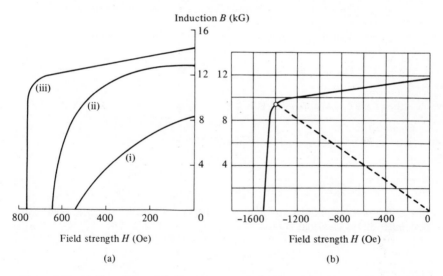

Figure 5.15. (a) The effect of field treatment and grain structure on the demagnetizing curves of an alloy with the composition 14% Ni, $7 \cdot 5$% Al, 25% Co, 3% Cu, $0 \cdot 7$% Nb, and balance % iron, cooled from 1250°C and tempered at 560–590°C: (i) Random orientation, no heat treatment in a magnetic field ('equi-axed'). $(BH)_{max} = 1 \cdot 75 \times 10^6$ gauss-oersted ($1 \cdot 75$ MGO). (ii) Random orientation of crystallites but alignment of precipitate by heat treatment in a field. $(BH)_{max} = 5 \cdot 4$ MGO. (iii) Columnar crystallites with heat treatment in a field. $(BH)_{max} = 8 \cdot 7$ MGO (Gould [15]).
(b) Demagnetizing curve for a single crystal of composition 35% Fe, $34 \cdot 8$% Co, $14 \cdot 9$% Ni, $7 \cdot 5$% Al, $2 \cdot 4$% Cu, and $5 \cdot 4$% Ti, heated for 10 min at 820°C, cooled at 1°C s^{-1} and tempered for 2 h at 650°C and 20 h at 585°C. $(BH)_{max}$ has the exceptional value of $13 \cdot 4$ MGO (de Vos [12]).

showed that this particular metallurgical effect was suppressed by the addition of small amounts of sulphur or selenium, and in consequence permitted the high coercivity and columnar structure to be obtained simultaneously [14]. Table 5.1, after Gould [15], indicates the extent of the improvement and also contains representative data for different types of material. More specific information can be obtained from commercial publications such as those issued by the Permanent Magnet Association of Sheffield, and a remarkably comprehensive table is given by Anselin [16].

It would be interesting to know how many of the technical achievements have been made on the grounds of theory, and how many from purely empirical development. With regard to the future it is not easy to predict the possibilities of further development due to the incomplete understanding of the interactions between the particles. The problem

Table 5.1.

(a) Representative properties of commercial Alnico-type alloys.

Type Examples	B_r (G)	H_c (Oe)	$(BH)_{max}$ (MGO)
Isotropic Alnico 2 (G.E.), Alnico (P.M.A.)	7000	550	1·6
Anisotropic Alnico 5 (G.E.), Alcomax III (P.M.A.)	12500	650	5·0
Columnar Ticonal 750 (Philips), Columax (P.M.A.)	13500	750	7·5

(b) Effect of titanium and niobium on properties of anisotropic (equi-axed) and columnar alloys (residual composition approximately 28% Co, 15% Ni, 7% Al, 3% Cu, balance Fe), after Gould [15].

Ti	Nb	S	Equi-axed			Columnar		
			B_r (G)	$(BH)_{max}$ (MGO)	H_c (Oe)	B_r (G)	$(BH)_{max}$ (MGO)	H_c (Oe)
1·0	0	0·18	11900	4·4	700	13100	6·7	805
3·7	2	0·21	9000	3·5	1090	11300	9·2	1360
4·0	2	0·17	8350	4·0	1195	10950	8·5	1350
5·5	0	0·22	8900	5·2	1475	10950	10·4	1590

illustrated by Figure 5.16 should certainly be amenable to numerical analysis. It is assumed that coherent rotation can be enforced by control of the size and shape of the particles, but when these are packed closely together the peak value of the shape anisotropy energy, indicated in (b), must be a function of s such that $W(s)_{max} \to 0$, as $s \to 0$, where $s = d/c$. The problem could be solved by computing the self-energies and inter-action energies of the sheets of charge involved and the optimum value of s, for the greatest $(BH)_{max}$, found for this particular model.

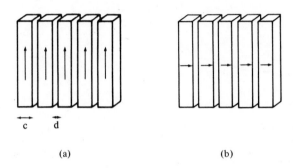

(a)   (b)

Figure 5.16. Model to illustrate the effect of magnetostatic interactions, (b) representing the peak in the shape anisotropy energy. As $d/c \to 0$, $B \to B_S$ but $H_c \to 0$, since the shape is effectively lost.

Beyond this further improvements could be made by utilizing crystal anisotropy, since the effects of shape and crystal anisotropies should be additive. Another intriguing possibility which has long been recognised but still does not appear to have been utilized in practice is shown in Figure 5.17. Part (a) of the figure shows an I/H loop for partially oxidized fine particles of cobalt which have been cooled from tempera-tures above the Néel point of the antiferromagnetic oxide in the presence of strong applied fields. The shifts in the loops give an increased H_c and $(BH)_{max}$ in one direction, which is all that is needed in practice, and if this could be enhanced or introduced into the alloys, it could be of real value. The source of this unidirectional uniaxial anisotropy is illustrated by Figure 5.17b: because of the super exchange between the surface metallic ions and the first layer of cations in the oxide layer, the ordering direction in the antiferromagnetic layer is tied to the direction of magnetization in the metal; the latter is aligned by the field applied while the ordering occurred. The fields applied during subsequent cycling of the magnetization have an inappreciable effect on the anti-ferromagnet, which thus maintains a unidirectional torque on the magnetization due to the exchange anisotropy [17].

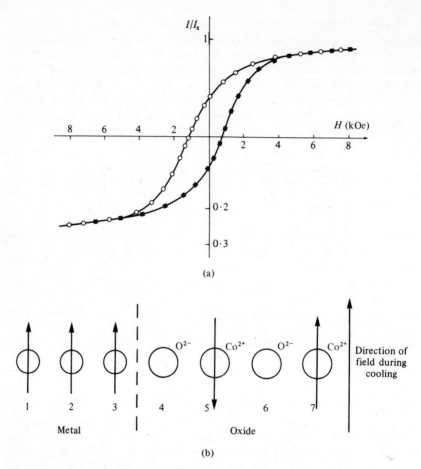

Figure 5.17. (a) A magnetization loop for oxidised cobalt cooled in a magnetic field, the shift along the *OH* axis indicating the presence of exchange anisotropy.

(b) Ions 1, 2, and 3 are metallic cobalt, 5 and 7 are Co^{2+}, and 4 and 6 are O^{2-}. The first three are mutually aligned by direct exchange, while 3 and 5, and 5 and 7 are aligned in an antiparallel manner by a super exchange mechanism via the intervening anions. Once the temperature has been reduced below the Néel temperature of CoO, the directions of 5 and 7 are 'frozen' in relation to moderate applied fields and a continuous torque is exerted on the metallic moments. In contrast to previous reports, it has been found that shifted loops can also be obtained for iron and nickel when very heavily oxidised.

5.7 COERCIVITIES ASSOCIATED WITH CRYSTAL ANISOTROPY

Curling and fanning do not apply to materials with high crystal anisotropy. In the absence of domain walls the coercivity of crystals with their easy axes aligned should be $2K/I_s$ for fields applied parallel to the easy axes, and should be independent of size.

The ratio $2K/I_s$ can be very large indeed. This is obviously so when it is realized that I_s can be made very small in a number of materials (for example in the orthoferrites $I_s \doteq 10$, $K \doteq 10^6 \, \text{erg cm}^{-3}$, predicting $H_c \doteq 100000$) but this alone does not form the basis for a general purpose permanent magnetic material. We may readily duplicate the calculations of the preceding section and find that

$$(BH)_{\text{max}} = 4\pi^2 I_s^2 \tag{5.34}$$

so long as

$$H_c = \frac{2K}{I_s} \geqslant 2\pi I_s \; ,$$

that is

$$K \geqslant \pi I_s^2 \; . \tag{5.35}$$

Thus, for a given value of K the magnetization should be as near as possible to the limit $(K/\pi)^{1/2}$ to generate the highest $(BH)_{\text{max}}$. (Alternatively, if we observe that the coercivity is a structure-sensitive property, we may note that for a material with given I_s, increasing the coercivity beyond $2\pi I_s$ does not lead to an improvement in $(BH)_{\text{max}}$ as conventionally defined, although it may be of value for particular applications). Some values of $2K/I_s$ have been given in Chapter 2 and Table 2.2 gives representative commercial values for technically useful materials.

In practice it must be stressed that coercivities close to $2K/I_s$ are achieved only by paying very careful attention to structural details, particularly to particle or grain size: despite the opening paragraph it is an observed fact that the coercivity *is* a function of grain size, and follows a form originally demonstrated for MnBi by Guillaud [18] as shown in Figure 5.18. The most common permanent magnet materials in this present category are the hexagonal uniaxial ferrites of barium or strontium, or more generally the compositions $(M_1O)_{1-x}(M_2O)_x k(Fe_2O_3)$ in which M_1 and M_2 are combinations of barium, strontium, lead and calcium. (Barium ferrite is $BaO.6Fe_2O_3$.) High coercivities are only obtained in sintered specimens if the powders are very finely milled before pressing and sintering in order to give an eventual grain size of about 10^{-4} cm, and even then the values are usually well below $(2K/I_s) - 4\pi I_s$, particularly for grain-oriented specimens. Typical B/H loops are shown in Figure 5.19. The grain-oriented specimens are prepared by compacting the ferrite grains in an aligning magnetic field before sintering, which then itself further improves the alignment.

The quantity $4\pi I_s$ included above represents the maximum shape anisotropy which opposes the crystal anisotropy in these materials, since the crystallites tend to be plate-shaped with the easy axes normal to the greater dimensions.

Dilute samples of barium ferrite with very fine grains, about 10^{-5} cm in diameter, do have coercivities near to the theoretical values, but of

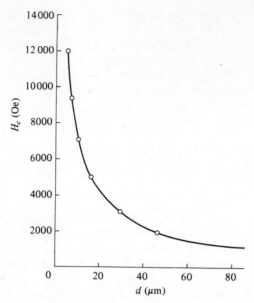

Figure 5.18. Observed variation of coercivity with particle size for manganese bismuthide (Guilland [18]).

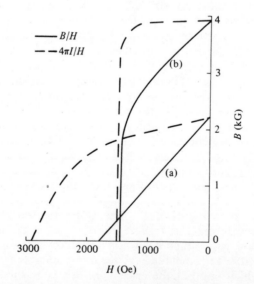

Figure 5.19. The effect of grain orientation on the demagnetizing curves of barium ferrite:
  (a) Random orientation (isotropic), $(BH)_{max} = 1 \cdot 04$ MGO.
  (b) Grains having easy axes aligned (anisotropic), $(BH)_{max} = 2 \cdot 93$ MGO.
It is assumed that the measurements are always made with fields applied parallel to the common easy axis. Note that $_BH_c = {}_IH_c$ for the aligned material, but the two coercivities differ greatly for the isotropic material (Gould [15]).

course the $(BH)_{max}$ values are not high [19]. In the course of sintering the density and orientation, and consequently the remanence, are improved but the simultaneous grain growth is associated with a decrease in coercivity. This grain growth has been inhibited, without affecting the improvement in remanence, by adding SiO_2 (in the case of strontium ferrite): the effect of the SiO_2 is associated with its melting and wetting the ferrite particles [20].

A recently developed class of alloys of special interest has the general formula Co_5R, where R is a rare-earth metal such as samarium. These alloys have anisotropy fields of around 200 000 Oe and $4\pi I_s \doteqdot 10 000$, which predicts possible maximum energy products of 40 MGO (mega-gauss oersted). The cast alloys are brittle and can be ground to fine powders, and it is found that coercivities of 10 000 Oe can be obtained in fine particle specimens. This again leaves a large discrepancy with H_K, which in this case may be connected with the crudity of the method of preparing the powders: annealing gives a certain recovery of H_c [21].

Reference to these remarkable values raises the question of how high one might require coercivities to be. When magnets are used in situations in which they are exposed to large fields of external origin, there is no real upper limit. So far as the production of fields is concerned, it must be remembered that the least favourable geometry is a flat disc or sheet magnetized normal to its surface, when the material is subject to demagnetizing fields approaching $4\pi I_s$. The limit itself represents a rather unique situation, since it is only necessary to add vectorially the uniform fields arising from two sheets of opposite charge in close proximity to see that the external field is just zero.

Obviously the important problem arising out of the above empirical account concerns the apparently anomalous variation of coercivity with particle size. It is known that if large crystals of barium ferrite, for example, are saturated and then removed from the applied field, domains form in the specimen's demagnetizing fields to effect substantial demagnetization: the coercivity is then only the relatively low field required to move the walls past any inhomogeneities in the structure (see Chapter 6). Thus, it would appear that the problem is equivalent to finding a size-dependence of the nucleation fields. Before exploring this, however, it is interesting to determine what behaviour is expected if the nucleation fields are neglected altogether.

5.8 SIZE-DEPENDENT EFFECTS WITH NEGLIGIBLE NUCLEATION FIELDS

It is assumed that if the nucleation field $H_n \to 0$ and the crystal contains no gross imperfections, then on the removal of a saturating field the magnetization distribution taken up will be that which corresponds to the minimum energy. If this is a multidomain state, with equal volumes of material magnetized parallel and antiparallel to the field direction, or $I_s v^+ = I_s v^-$, then $H_c \to 0$. In the presence of imperfections H_c is equal to H_w, the 'wall coercive field', which will be assumed to be negligible.

For a uniaxial crystal we account for the magnetostatic energy E_m and 180° domain wall energy E_γ only (no changes in E_K). The former is proportional to the volume, $E_m \propto a^3$ for a cube with side a, while $E_\gamma \propto a^2$ for a single wall bisecting the cube. As the dimensions are reduced it is to be expected that a stage will be reached at which the introduction of a domain wall into a single-domain crystal ceases to reduce $(E_m + E_\gamma)$: the point at which this occurs defines the critical size for the single-domain state in zero field.

An approximate value for the critical radius of a spherical uniaxial crystal is readily found on the assumption that a central wall halves the magnetostatic energy. The energy of the single domain state is

$$E_s = \tfrac{1}{2} N I_s^2 \times (\text{volume}) = \tfrac{1}{2} \times \tfrac{4}{3}\pi I_s^2 \times \tfrac{4}{3}\pi r^3 ,$$

and for the twin-domain state

$$E_t = \tfrac{1}{2} W_s + \pi r^2 \gamma .$$

Equating these expressions gives

$$r_c = \frac{9\gamma}{4\pi I_s^2} . \tag{5.36}$$

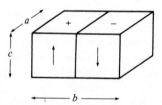

Figure 5.20. Model used for calculating critical sizes for the single-domain equilibrium structure. According to calculation the domain wall always runs parallel to the shorter side as shown, assuming $b > a$.

A more comprehensive and precise treatment can be given for rectangular blocks with dimensions $a \times b \times c$, with $I_s \| c$ as shown in Figure 5.20 [22]. W_s is calculated as the sum of the two (identical) self energies of the sheets of charge less the interaction energy of the two, and can be expressed as

$$E_s = S\left(\frac{b}{a}, \frac{c}{a}\right) a^3 I_s^2$$

while the magnetostatic energy of the twin-domain configuration is similarly given by another function $T(b/a, c/a)$. If the dimensions at the critical size are represented by ka, kb, kc then

$$Sk^3a^3I_s^2 = Tk^3a^3I_s^2 + k^2\gamma ac .\qquad(5.37)$$

If the unit of length chosen is γ/I_s^2 (which is dimensionally correct) the results apply to any material. Results of such calculations, for a range of values of b/a, are shown in Figure 5.21. For a cube, for example, the critical reduced dimension is $a = 0\cdot95$ as compared with the sphere diameter of $9/2\pi = 1\cdot43$. It can also be seen that as the elongation increases, the total volume at the critical size rises.

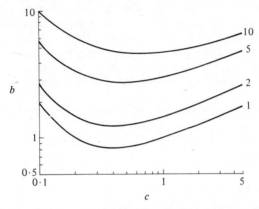

Figure 5.21. The effect of shape on the critical size of an isolated crystal, as shown in Figure 5.20: the numbers on the curves represent values of b/a (reduced dimensions).

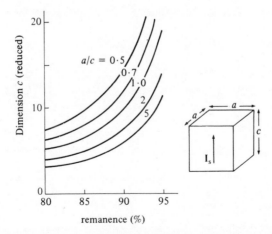

Figure 5.22. The critical size of interacting close-packed crystallites drawn as a function of shape (a/c) and of the degree of orientation, represented by the calculated remanence (reduced dimensions).

The above discussion refers to isolated crystals. An important type of practical material consists of polycrystalline specimens in which the grains have their easy axes oriented nearly parallel to a common axis. For perfect orientation it is expected that the separate identity of the grains should be lost. Otherwise, a pole density, which is dependent on the degree of misorientation, will be developed at the grain boundaries; the critical size can be computed from this by again equating the energies when each grain consists of a single domain or two domains. It is convenient to relate the results to a readily measured property which is also a function of the degree of alignment, that is to the remanence I_r/I_s.

Figure 5.22 shows the rapid increase in the critical size at high values of the remanence for crystallites of the shape indicated, and again demonstrates an increase in critical size with the degree of elongation of the individual crystallites. An experimental value of the critical size for oriented barium ferrite has been obtained from domain studies on thermally demagnetized specimens. This is about 2×10^{-4} cm: the grains were foreshortened with $a/c \doteqdot 1 \cdot 5$, so the critical size in reduced dimensions is $a = b = 5 \cdot 6$, $c = 3 \cdot 8$; if we use a domain wall energy of 5 erg cm^{-2}, $I_s = 380$. The experimental value corresponds to a remanence of 80%, which is rather low, and the problem cannot be said to be completely solved.

5.8.1 The effect of applied fields

Recalling that the demagnetizing energy of a cube is $\frac{1}{2} \times \frac{4}{3}\pi I_s^2$, as for a sphere, the energy of a cubic single domain particle in a reverse field H is

$$E_s = \tfrac{1}{2} \times \tfrac{4}{3}\pi a^3 I_s^2 + Ha^3 I_s \,. \tag{5.38}$$

This is plotted as line (1) in Figure 5.23. The energy of a twin-domain

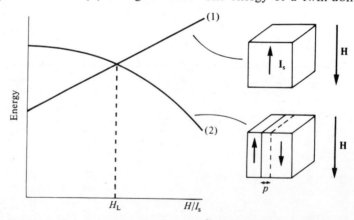

Figure 5.23. The variation of the energy with applied field for an isolated crystal, of less than the critical size in the absence of a field, in the single domain state (1) and in the twin domain state (2). Above a (reduced) field value H_L, the twin domain state has the lower energy.

particle is readily calculable if it is assumed that the domain wall is displaced by the field through a distance p as

$$E_t = T(p)a^3 I_s^2 + a^2 \gamma - 2pHa^3 I_s . \qquad (5.39)$$

The first term represents the magnetostatic energy, and $T(p)$ is a numerically calculable function of the wall displacement; the second term is the wall energy; and the last term is the energy of the magnetization in the applied field. When p itself is calculated, E_t can be derived and plotted as line (2) in Figure 5.23. Here it has been assumed that in the absence of a field the crystal is below the critical size ($E_s < E_t$ at $H = 0$), but it is seen that the two curves cross at a field H_L above which the twin-domain state has the lower energy. From a series of figures such as this it is possible to derive the variation of the cube side (expressed as a fraction of the zero-field critical value) with the reduced field (H_L/I_s), for which the two energies are equal as in Figure 5.24. Now for the magnetization to be reversed by wall movement the wall must, at some stage, stretch across the crystal, and so these values of H_L/I_s represent lower bounds to the coercivity.

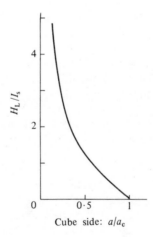

Figure 5.24. The size dependence of H_L (see Figure 5.23) for a uniaxial crystal of cubic form (reduced dimension: a_c is the critical size in zero field).

It is only if this latter calculation is appropriate that the critical size calculations have any relevance at all to coercivity. (They certainly are important in other respects.) The form of Figure 5.24 is about right (see Figure 5.18) but the effect portrayed only applies for $a < a_c$, whereas in practice the variation in H_c appears to spread to values above a_c. This is still an active field of research.

5.9 DOMAIN NUCLEATION

A perfect spherical crystal, with high K/I_s, should remain uniformly magnetized at remanence, since the demagnetizing fields are uniform. When reverse fields are applied the magnetization should be entirely unaffected until $H = H_K$, whatever the size. In fact, reverse domains nucleate in large crystals in fields well below H_K, even though the initial formation of a reverse domain requires the rotation of the magnetization in a certain volume of material against forces of anisotropy and exchange.

Clearly nucleation must be critically dependent on internal defects or on details of shape. Nucleation would appear to be favoured by particularly high local values of the internal stray fields, and indeed the formulae developed in Chapter 1 show that the field components normal to the magnetization can approach infinity. However the analyses given below (section 5.9.1) indicate that in practice the effectiveness of stray fields appears to be very limited.

Defects of some sort seem to be indicated as the real source of low nucleation fields, but it is still far from clear just how these are effective, or indeed what form of defects are implicated. The author's personal view favours dislocation clusters in barium ferrite, but this cannot yet be absolutely justified. One recent result of considerable interest concerns europium orthoferrite which, having a very high value of K/I_s, should have very high nucleation fields. In strain-free crystals, in fact, these are very high, but they are greatly reduced by polishing the crystal surfaces. This is natural, but the more finely the surfaces are polished the lower the nucleation fields become (down to 10 Oe after fine diamond polishing). Two other observations (made by Constantin Tanasoiu in the author's laboratory) indicate the remarkable complexity of the problem: as shown in Figure 5.25, the nucleation fields may be a function of the previously applied 'saturating field', and they may have a unidirectional character by being greater as measured in one direction along the easy axis. It would be of considerable value to know the precise structure and composition of the layer of modified material produced by the polishing action. In single crystals the dependence of nucleation fields on the magnitude of the saturating field was first demonstrated for yttrium orthoferrite [23].

Assuming the effectiveness of some arbitrary defects one can readily produce a plot of H_c versus size of almost any reasonable form, simply on the assumption that the larger grains are more likely to contain substantial defects. In case of difficulty one can always assume a distribution of defects with a range of associated nucleation fields. It is hardly necessary to comment that the general situation is far from satisfactory.

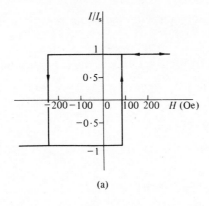

(a)

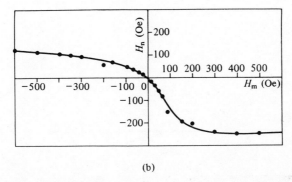

(b)

Figure 5.25. (a) An asymmetrical magnetization loop for a small crystal of europium orthoferrite, $EuFeO_3$. Magnetization reversal occurs by the nucleation of domain walls, which are driven through the crystal to cause saturation, that is H_n is greater than the field required for saturation. H_n clearly depends upon the direction of the applied field.

(b) The dependence of H_n on the direction of switching, and on the magnitude of the field applied before switching (H_m).

5.9.1 Magnetostatic analysis of domain nucleation and growth

Two specific effects have been suggested to give rise to a relation on purely magnetostatic grounds between nucleation fields and crystal size. It has been shown in Chapter 1 that a plane sheet of charge generates very high field components parallel to the surface of the sheet; indeed, these fields tend to infinity at points which approach the edge of the sheet, while the components normal to the sheet never rise above $2\pi I_s$. Thus, it was natural to suppose [24] that these lateral fields could assist the reversal of the magnetization in a certain volume of the material close to a sharp corner or edge of a crystal surface, particularly since the magnitude of the lateral field at any point is itself a function of the area of the surface concerned.

If we consider a crystal with square cross-section, $2a \times 2a$, normal to the magnetization direction (Figure 5.26a) the field parallel to OX at the point $(0,0,c)$ due to the elementary charge $\sigma\, dx\, dy$ at $(x,y,0)$ is $x\sigma\, dx\, dy/(x^2+y^2+c^2)^{3/2}$ and on integrating over the whole square

$$H_L = 2\sigma\left[\sinh^{-1}\frac{a}{c} - \sinh^{-1}\frac{a}{(4a^2+c^2)^{1/2}}\right] . \qquad (5.40)$$

Similarly, at the point $(c,0,c)$

$$H_L = 2\sigma\left[\sinh^{-1}\frac{a}{c\sqrt{2}} - \sinh^{-1}\frac{a}{(4a^2+c^2)^{1/2}}\right] . \qquad (5.41)$$

and for $c \ll a$ this becomes (with $\sigma = I_s$)

$$H_L = 2I_s\log\frac{a}{c} . \qquad (5.42)$$

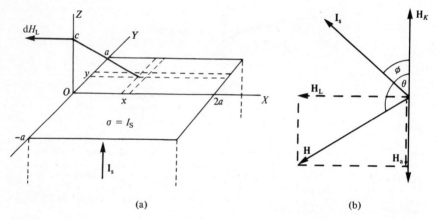

(a)   (b)

Figure 5.26. (a) Diagram for the calculation of the fields arising from the surface of a uniformly magnetized crystal. (b) The combined effects of the stray field H_L and applied field H_a on the direction of magnetization.

Now consider that the magnetization is caused to rotate through an angle ϕ by the joint effect of H_L and a field H_a applied parallel to the easy axis, the resultant of the two having magnitude H, equal to $(H_a^2+H_L^2)^{1/2}$, and orientation θ as shown in Figure 5.26b. The net torque on the magnetization is

$$\Gamma = HI_s \sin(\theta-\phi) - K\sin 2\phi . \qquad (5.43)$$

The field will cause the magnetization to rotate past $\phi = \pi/2$, if the minimum value of Γ is zero, that is if $dT/d\phi = 0$ and $\Gamma = 0$ simultaneously. Solving the two resulting equations gives

$$H_L^{2/3}+H_a^{2/3} = \left(\frac{2K}{I_s}\right)^{2/3} . \qquad (5.44)$$

This value of H_a is the lower bound to the nucleation field, H_n, and if we use the value of H_L given in Equation (5.42), then

$$H_n^{2/3} = \left(\frac{2K}{I_s}\right)^{2/3} - \left(2I_s \log\frac{a}{c}\right)^{2/3} . \tag{5.45}$$

Thus, for any arbitrarily fixed value of c, the nucleation field will be seen to vary with a, but only logarithmically (that is very slowly). Moreover, the calculated value can only be fitted to Guillaud's data by putting $c = 10^{-10}$ cm. Apart from being below atomic dimensions this value clearly invalidates the whole approach since the magnetization must, in fact, rotate throughout a volume of material equivalent at least to the normal domain wall width or the exchange energy will be very high. (The omission of exchange energy was the reason for calling the calculated H_n the lower bound.)

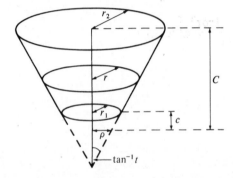

Figure 5.27. Representation of a surface pit as a truncated cone, for calculation of the field at the point P distance c below the base of the pit.

In a second approach to this problem [25] it has been noted that the field at the base of a conical etch pit is given by

$$H = NI_s \log\frac{r_2}{r_1} . \tag{5.46}$$

with r_2 and r_1 as shown in Figure 5.27 and N a factor which depends on the angle of the cone. It was assumed that the larger crystals are more likely to contain such surface defects and this was suggested to lead to a size-dependence of the nucleation fields. Again, however, it is necessary to investigate the spatial distribution of the fields in order to estimate their true significance in reinforcing the applied fields. The field at a point P, distance c from the base of the pit, is given by

$$\int_{r_1}^{r_2} \frac{dH}{dr} dr ,$$

where

$$dH = \frac{2\pi r I_s (z-h)\,dr}{[r^2+(z-h)^2]^{3/2}}\,,$$

and the field due to the base of the pit and the surrounding surface is

$$\int_0^{r_1} \frac{2\pi r c I_s\,dr}{(r^2+c^2)^{3/2}} + 2\pi I_s C \int_{r_2}^\infty \frac{r\,dr}{(r^2+C^2)^{3/2}}$$

(see Figure 5.27). Evaluation of these integrals gives the magnitude of the field (directed along the axis of the cone) as

$$
\begin{aligned}
H = {}& \frac{2\pi\sigma t^2}{(1+t^2)^{3/2}}\Big[\log\{r(1+t^2)^{1/2}-\rho+(1+t^2)^{1/4}[(1+t^2)r^2-2r\rho+\rho^2]^{1/2}\}\Big]_{r_1}^{r_2} \\[4pt]
& -\frac{2\pi\sigma t^2\rho(1-t^2)}{(1+t^2)^2}\Big[\{r^2(1+t^2)-2r\rho+\rho^2\}^{-1/2}\Big]_{r_1}^{r_2} \\[4pt]
& -\frac{4\pi\sigma t^2}{(1+t^2)^{3/2}}\Big[\tanh\Big(r-\frac{\rho}{1+t^2}\Big)\Big]_{r_1}^{r_2} \\[4pt]
& +\frac{2\pi\sigma(r_2-\rho)}{t}\Big[r_2^2+\Big(\frac{r_2-\rho}{t}\Big)^2\Big]^{-1/2} \\[4pt]
& -\frac{2\pi\sigma(r_1-\rho)}{t}\Big\{\Big[r_1^2+\Big(\frac{r_1-\rho}{t}\Big)^2\Big]^{-1/2}-\Big(\frac{t}{r_1-\rho}\Big)\Big\} \qquad (5.47)
\end{aligned}
$$

where $\rho = r_1 - ct$ (see Figure 5.27). This expression for H increases with increasing r_2 and with decreasing r_1 and c. On taking extreme values of r_2 equal to the crystal diameter, say 10^{-4} cm, and r_1 equal to the unit cell size, one still only obtains values of $H = 20 I_s$ for $c = 10^{-7}$ cm; $H = 10 I_s$ for $c = 10^{-5}$ cm. For barium ferrite the nucleation field so predicted can scarcely be less than $(2K/I_s) - 20 I_s = 8000$ Oe. Thus, the effectiveness of the fields derived in this way is again very limited, and this approach would also appear to be invalidated.

Although the results of these two analyses are purely negative their reproduction does seem worth while, since the calculations themselves are of interest and may have other useful applications, and the subject of nucleation is of such basic importance.

There are still more 'negatives' in this topic. Aharoni [26] considered the effects of thermal fluctuations on the magnetization direction in a small region, and showed that this could not explain the observed nucleation effects.

Shtol'ts prepared particles of MnBi which had a similar variation of coercivity with size to those studied by Guillaud, and noted that particles with $H_c \doteqdot 3000$ Oe contained single reverse domains at remanence [27]. Clearly nucleation was not important in these particles, and the coercivity was controlled by a resistance to domain growth. Again this is very difficult to understand. The situation is amenable to magnetostatic analysis, but the results are far from illuminating.

5.10 THE EFFECT OF DAMPING ON ROTATIONAL SWITCHING AND SUSCEPTIBILITY

It was shown in Chapter 2 that in the absence of damping (when there is no means of transferring energy from the magnetization, or the atomic moments, to the surroundings), an applied static field caused precession only, and there is no change in the component of the magnetization in the field direction. In the present chapter, on the contrary, it has been assumed that equilibrium has always been achieved, so that in the absence of net restoring torques the magnetization always tends to follow the field direction. The damping, which has been implied to permit this behaviour, must also lead to the conclusion that the magnetization cannot follow the field direction instantaneously since this would involve a power loss, or rate of dissipation of energy, approaching infinity. The two important practical consequences of this conclusion are that the switching time, or time for reversal of the magnetization, must always be finite, and that the susceptibility must be frequency dependent.

Without discussing its origin, damping can be introduced formally by the addition of a relaxation term to the equation of motion as originally suggested by Landau and Lifshitz [28], to give

$$\frac{d\mathbf{I}}{dt} = -\gamma(\mathbf{I} \times \mathbf{H}) + \frac{\lambda}{I^2}[\mathbf{I} \times (\mathbf{I} \times \mathbf{H})] \; , \qquad (5.48)$$

where λ is the relaxation frequency or damping constant, and γ the magnetomechanical ratio defined in Chapter 2. Gilbert [29] suggested that the relaxation term should be proportional to the rate of change of the magnetization, as in

$$\frac{d\mathbf{I}}{dt} = -\frac{\gamma^*}{1+\alpha^2}(\mathbf{I} \times \mathbf{H}) + \frac{\alpha}{I}\left(\mathbf{I} \times \frac{d\mathbf{I}}{dt}\right) \quad . \qquad (5.49)$$

This equation can be rearranged after some manipulation (multiplying vectorially by $\mathbf{I}$) as

$$\frac{d\mathbf{I}}{dt} = \frac{-\gamma^*}{1+\alpha^2}(\mathbf{I} \times \mathbf{H}) + \frac{\gamma^*}{1+\alpha^2}\frac{\alpha}{I}[\mathbf{I} \times (\mathbf{I} \times \mathbf{H})] \; . \qquad (5.50)$$

Equation (5.50) is identical to the Landau–Lifshitz equation if

$$\gamma^* = \gamma(1+\alpha^2)$$

and

$$\frac{\lambda}{I} = \frac{\gamma^*\alpha}{1+\alpha^2} \; .$$

If $\alpha^2 \ll 1$ we have $\gamma^* \doteq \gamma$ and the two equations are equivalent with

$$\alpha = \frac{\lambda}{I\gamma} . \tag{5.51}$$

For most resonance experiments (see below) the condition $\alpha^2 \ll 1$ is applicable and it appears customary to use the Landau–Lifshitz equation. Both equations can be written

$$\frac{d\mathbf{I}}{dt} = -\gamma\,[\mathbf{I} \times (\mathbf{H} - \mathbf{H}_\lambda)] , \tag{5.52}$$

where $\mathbf{H}_\lambda$ is the effective field representing the damping forces. According to Gilbert this field is given by

$$\mathbf{H}_\lambda = \frac{\alpha}{\gamma^* I}\frac{d\mathbf{I}}{dt} .$$

5.10.1 Precession and switching

Figures 5.28a and b show the response of the magnetization to an applied field in two obviously limiting cases. Figure 5.28a represents pure precession, for λ (or α) equal to zero. If the damping is very heavy, a direct but slow approach of the magnetization direction towards that of the applied field is expected (Figure 5.26b). In general, however, the motion of the magnetization vector will combine both these features; that is to say it will perform a spiral and gradually approach its equilibrium direction (the direction of the field) as shown in Figure 5.28c.

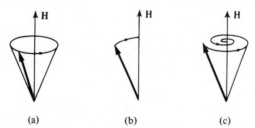

Figure 5.28. Representation of the response of the magnetization to an applied field: (a) With no damping; a simple precession at a frequency $\omega_0 = \gamma H$. (b) With very heavy damping; a relatively slow rotation directly towards the field direction. (c) A more general case.

This motion can be described by equations for the time-dependence of the three components of the magnetization shown, which are solutions of the Landau–Lifshitz equation:

$$I_x = I_i e^{-t/\tau}\cos\omega_0 t \qquad\qquad I_y = I_i e^{-t/\tau}\sin\omega_0 t$$

$$I_z = I_s\left[1 - \left(\frac{I_i}{I_s}\right)^2 \exp\left(-\frac{2t}{\tau}\right)\right]^{\frac12} \tag{5.53}$$

ω_0 is the frequency of free precession (equal to γH) and τ is the relaxation time or characteristic time for the motion (equal to $I_s/\gamma H$). I_i is the initial value of the projection of the magnetization vector on the plane OXY, namely at $t = 0$.

For some particular value of λ (or α) it may be expected that the magnetization will describe an optimum path and give the minimum value of the switching time. This is an elaborate problem, but gives the simple result that critical damping is represented by [30]

$$\lambda = \gamma I_s , \tag{5.54}$$

in which case the minimum switching time is

$$\tau_{min} = \frac{2}{\gamma H} . \tag{5.55}$$

This represents very fast switching. For electron spin $g = 2 \cdot 0023$ and the magnetomechanical ratio $\gamma = 1 \cdot 7609 \times 10^7 \, s^{-1} \, Oe^{-1}$, so that for an applied field of only 1 Oe, $\tau_{min} = 0 \cdot 1 \, \mu s = 100$ ns.

τ_{min} can be even less than this, if the damping is sub-critical and the precession partly controlled by the effects of the shape of the specimen. For the flat disc, or film, shown in Figure 5.29, in which the magnetization is to be rotated through $90°$, true precession about the applied field direction would involve the creation of a large component of the magnetization across the surface. The magnetization is thus caused to precess largely within the plane of the film [30], and in the equation for precession the applied field can be replaced by the effective field due to shape anisotropy, which in this case is $4\pi I_s$. (This is not an implication that actual demagnetizing fields of this magnitude arise, which would clearly be a contradiction.) If the oscillatory part of the magnetization change is ignored, and the damping is very low, we now have a switching time which is the reciprocal of the precessional frequency:

$$\tau = \frac{1}{f} = \frac{2\pi}{\omega} = \frac{2\pi}{\gamma H_e} = \frac{1}{2\gamma I_s} . \tag{5.56}$$

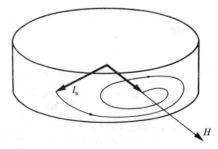

Figure 5.29. The modification of the motion shown in Figure 5.28c by the demagnetizing effects in a thin specimen.

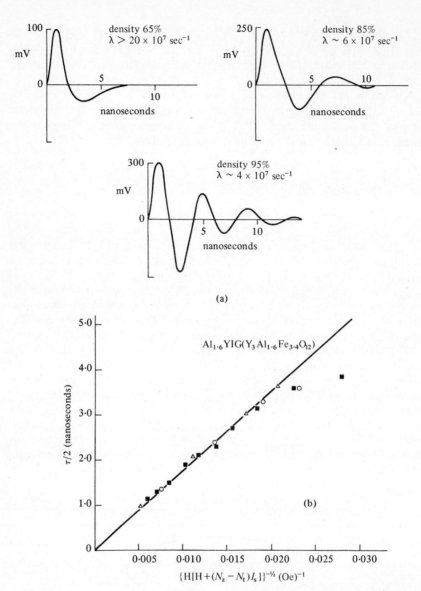

(a)

(b)

Figure 5.30. (a) Damped precession of the magnetization in discs of aluminium substituted YIG after the application of pulsed fields, as indicated by the voltage in a conductor close to the specimen. Such fields effectively rotate a saturating bias field through a small angle in less than 1 nanosecond. The reciprocal of the relaxation time, obtained from the decrement of the signal, is proportional to the porosity up to 20% porosity.
(b) Linear relation between the precession half period, $\tau/2$, and $1/H_{\text{effective}}$ corresponding to $\omega = \gamma\{H[H+(N_z-N_t)I_s]\}^{\frac{1}{2}}$ where N_z and N_t are demagnetizing factors respectively normal and parallel to the plane of a disc. Three specimens, each with a different value of I_s, were produced by quenching, and examined in a range of applied fields. (D.J.Craik, A.D.Kitto, and D.P.Ross, to be published.)

If we take the low value of $I_s \doteqdot 100$, we get $\tau \doteqdot 1$ ns. To appreciate just how fast this is, it may be recalled that the velocity of light is 3×10^{10} cm s^{-1}, so that 10^{-9} s is the time taken for light to travel about one foot.

Experimentally, current pulses with a very short rise time (less than one nanosecond) can be produced by using a mercury-wetted relay. If these pulses are passed along a conductor carefully positioned with respect to an earthed sheet, that is a strip line, in close proximity to a thin slice of a high-resistivity ferrite, the magnetization can be shown to behave in the manner illustrated in Figure 5.28. This may be seen by observing the voltage pulses induced in a second conductor orthogonal with the drive strip: since this conductor intersects the stray field from the ferrite, the induced voltage is proportional to the rate of change of the magnetization. Typical output pulses are shown in Figure 5.30 for the materials described in the caption. Sampling oscilloscopes can be used to resolve time intervals of less than 10^{-9} s.

Possible sources of damping are eddy-currents in metal films above a certain thickness and relatively high conductivity ferrites: electron diffusions in oxides containing divalent and trivalent ions in neighbouring lattice sites, and the transfer of energy to lattice vibrations via spin waves. These will be discussed further below and damped domain wall motion will be treated in Chapter 6.

5.10.2 Permeability or susceptibility spectra: losses

The time delay between the application of a field and the full response of the magnetization must have important consequences when alternating fields are applied. At vanishingly low frequencies the magnetization will remain in phase with the field whatever the damping (so long as it is finite) and the initial susceptibility will be as calculated in the preceding sections. More generally the lag can be represented by a phase angle δ, that is to say, if the a.c. field is

$$H(t) = H_1 \cos \omega t ,$$

then

$$I(t) = I_1 \cos(\omega t - \delta)$$
$$= I_1 \cos \omega t \cos \delta + I_1 \sin \omega t \sin \delta . \qquad (5.57)$$

This equation can be rewritten

$$I(t) = H_1 \chi' \cos \omega t + H_1 \chi'' \sin \omega t \qquad (5.58)$$

by defining

$$\chi' = \frac{I_1 \cos \delta}{H_1} \quad \text{and} \quad \chi'' = \frac{I_1 \sin \delta}{H_1} ,$$

so that $\chi''/\chi' = \tan \delta$. From the definition of δ, $I_1 \cos \delta$ is the component of the induced magnetization which is parallel to H_1, and so

χ' is the 'in phase' susceptibility and similarly χ'' is the 'out of phase' susceptibility.

Using complex number notation the alternating field is $H_1 e^{i\omega t}$, the magnetization is $\chi H_1 e^{i\omega t}$ and one obtains the above results if χ is taken to be complex:

$$\chi = \chi' - i\chi'' = |\chi| e^{-i\delta} \qquad (5.59)$$

(using $e^{i\omega t} = \cos\omega t + i\sin\omega t$). Thus, χ' and χ'' are known as the real and imaginary parts of the complex initial susceptibility.

The corresponding permeabilities, $\mu' = 1 + 4\pi\chi'$, $\mu'' = 1 + 4\pi\chi''$ may readily be measured. If a core of magnetic material, with diameter D and cross-sectional area A, is wound by N turns and the characteristics measured by bridge techniques, it is found to behave as an impedance

$$Z = R + i\omega L_s \,,$$

with the equivalent resistance

$$R = \omega \frac{N^2 A}{D} \mu'' \times 10^8 \text{ ohms} \,,$$

and self inductance

$$L_s = \frac{N^2 A}{D} \mu' \times 10^{-8} \text{ henries} \,.$$

Clearly, at very low frequencies $\delta \to 0$, $\chi'' \to 0$ and $\chi = \chi' = \chi_0$, the value calculated for static behaviour. As ω is increased δ increases, χ' falls and χ'' rises. The variation of the permeability (susceptibility) with frequency is known as the permeability (susceptibility) spectrum. A regular fall in χ' is said to represent a relaxation, as distinct from resonant behaviour (discussed in the next section). The exact way in which χ' and χ'' vary depends upon the relaxation or damping mechanisms involved. If these mechanisms can be characterized by a single time constant, that is if there is just one particular source of damping such as eddy currents or electron diffusion, the rate of change of the magnetization can be expected to be proportional to its deviation from its equilibrium position, whence we may write:

$$\frac{dI}{dt} = \frac{1}{\tau}(I_\infty - I) \qquad (5.60)$$

where I_∞ represents the component of the magnetization in the field direction when equilibrium is attained. On integration, after separating the variables:

$$I = \chi_0 \left[1 - \exp\left(-\frac{t}{\tau}\right) \right] H = \left[1 - \exp\left(-\frac{t}{\tau}\right) \right] I_\infty \qquad (5.61)$$

since $\chi_0 = I_\infty/H$, which means there is an exponential approach to the equilibrium value. Putting $H = H_1 e^{i\omega t}$, then

$$I = \frac{\chi H}{1 + i\omega t}$$

and

$$\mu' = 1 + \frac{4\pi\chi_0}{1 + \omega^2\tau^2} \qquad \mu'' = \frac{4\pi\chi_0\omega\tau}{1 + \omega^2\tau^2} . \tag{5.62}$$

The form of the two normalized functions of frequency is shown in Figure 5.31. Results corresponding approximately to this behaviour are obtained for magnetite [31] (as shown in Figure 5.31b), for spinel ferrites containing copper [32], and generally for any spinel or garnet ferrites which contain both divalent and trivalent ions on B sites [33]. The effective interchange of ferric and ferrous ions does not require ionic diffusion but only electron diffusion, $Fe^{3+} + e \rightleftharpoons Fe^{2+}$. The driving force for the diffusion is connected with an induced anisotropy, namely the directional dependence of the energy of the magnetization on its orientation with respect to certain configurations of ferrous and ferric ions. This electron diffusion has a finite relaxation time which gives rise to the lag or phase difference. Since the diffusion is thermally activated, the relaxation time is a function of temperature and the activation energy can be measured from the slopes of plots of the logarithm of the relaxation time against the reciprocal temperature ($\tau \propto e^{E/kT}$). The same processes provide a mechanism for electrical conductivity and materials such as manganese–zinc–ferrous ferrite, with relatively low resistivity of $\rho \doteq 10^2$ ohm cm, give activation energies from magnetic measurements of the magnitude for direct electron diffusion, viz: $E \doteq 0 \cdot 1$ eV or less.

The association of high losses with a rapid fall of susceptibility is made by noting that the work done when the magnetization changes by dI is $dW = -H dI$. Thus the energy absorbed per second per unit volume of material is

$$\frac{dW}{dt} = \frac{\omega}{2\pi} \int H \, dI = \frac{\omega}{2\pi} \int H \, d(\chi'' H) ,$$

where the integration is made over one cycle to give

$$\frac{dW}{dt} = \frac{\omega}{2} \chi'' H_1^2 . \tag{5.63}$$

This power loss appears as heat in the sample. Due to the relation between δ and μ'', $\tan\delta$ is known as the *loss factor*. Equation (5.63) only gives the actual losses if the amplitude of the induced magnetization is low enough to be virtually reversible with negligible hysteresis loss. Thus, χ'' must be taken as the out of phase component of the *initial* susceptibility only.

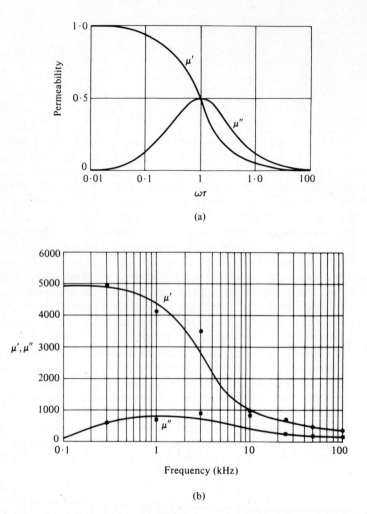

(a)

(b)

Figure 5.31. (a) The variation of the real and imaginary parts of the permeability, μ' and μ'' as forecast for relaxation with a single time constant [see Equation (5.62)]. (b) Relaxation spectra measured for magnetite, Fe_3O_4 (Galt [31]).

5.11 THE EFFECT OF RESONANCE ON LOSSES

While paramagnetic resonance was treated quantum mechanically in Chapter 3, it is legitimate to introduce ferromagnetic resonance in a classical manner since the spins are coupled by the exchange energy to give a macroscopic system.

In Figure 5.32 the magnetization is taken to be precessing, with $\omega_0 = \gamma H$ in a static field $\mathbf{H} \parallel OZ$; an alternating field is applied in the plane OXY, normal to $\mathbf{H}$. Any alternating field can be treated as the vector sum of two oppositely rotating fields of equal magnitude and angular frequency, as can easily be seen graphically. The counterclockwise field can be represented by the time dependence of its two components:

$$H_x = H_1 \cos \omega t$$
$$H_y = H_1 \sin \omega t \quad (H_z = 0) . \qquad (5.64)$$

If ω is far from ω_0, torques will be exerted which periodically tend to turn the magnetization towards or away from OZ, the average torque being zero. (The clockwise component can never be in phase with the precession and can be neglected.) If $\omega = \omega_0$, however, the rotating field remains in phase with the precessing magnetization and can exert a torque of constant sense and thus either increase or decrease the component along OZ. The possibility of absorbing energy from an alternating field with the same frequency as the precession is known as *resonance* [see also Equation (5.65)].

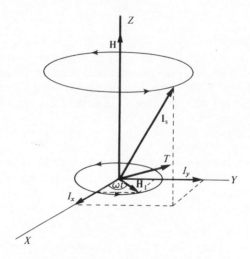

Figure 5.32. The magnetization vector $\mathbf{I}_s$ precessing in the static field $\mathbf{H}$, while a rotating field $\mathbf{H}_1$ is applied in a plane normal to $\mathbf{H}$.

If the components of the magnetization normal to OZ are I_x and I_y, the two complex magnetizations may be defined as

$$I_+ = I_x + iI_y ,$$
$$I_- = I_x - iI_y .$$

The components of the equation of motion of the magnetization, including the effects of the rotating field, are

$$\frac{dI_x}{dt} = \gamma(I_y H - I_z H_1 \sin \omega t) \ ,$$

$$\frac{dI_y}{dt} = \gamma(I_z H_1 \cos \omega t - I_x H) \ ,$$

$$\frac{dI_z}{dt} = \gamma(I_x H_1 \sin \omega t - I_y H_1 \cos \omega t) \ ,$$

and it follows that

$$\frac{dI_+}{dt} = i\gamma I_+ H + i\gamma I_z H_1 e^{i\omega t} \ ,$$

$$\frac{dI_-}{dt} = i\gamma I_- H - i\gamma I_z H_1 e^{-i\omega t} \ ,$$

$$\frac{dI_z}{dt} = \gamma(I_+ e^{-i\omega t} - I_- e^{i\omega t})H_1 \ ,$$

and also

$$I_\pm = \frac{\gamma H_1 I_z}{\gamma H + \omega} e^{\pm i\omega t} \tag{5.65}$$

If the susceptibility is defined as $I_\pm / H_1$, it clearly approaches infinity as $(\gamma H + \omega) \rightarrow 0$, that is as ω approaches the frequency of free precession, $-\gamma H$. This result provides an alternative definition of resonance, but in practice the susceptibility is always finite due to the inevitable presence of a certain amount of damping. Consequently, the losses do increase rapidly at resonance, but always remain finite.

Resonance in large static fields is not immediately relevant to permeability spectra, which are measured in alternating fields only, and it is necessary to reintroduce the concept of precession in effective fields. Thus, when the precession may be considered to occur in the anisotropy field, resonance occurs at

$$\omega_0 = \gamma H_K = \gamma \frac{4}{3} \frac{K}{I_s} \tag{5.66}$$

for negative K_1. When the effects of damping, and of the presence of domain walls, are taken into account, it is found that the permeability (μ_0) of low loss ferrites remains virtually constant up to ω_0 and then falls rapidly, the losses simultaneously rising (Figure 5.33).

Hence, two distinct effects of damping may be described. When the damping is high it leads to a relaxation at relatively low frequencies. When the damping is very low (as in ferrites with high resistivity, no appreciable ferrous ion content, etc.) the damping broadens the resonance loss peak and for high anisotropy fields the losses may not become appreciable until microwave frequencies are reached.

As a result there are two very important practical consequences, both of which demonstrate a degree of incompatibility between high values of μ_0 at low frequencies and a large range of useful frequencies over which μ_0 remains constant. The first is that the lowest values of the anisotropy (in ferrites) appear to be obtained for manganese–zinc ferrites, which have relatively low resistivity. The second, more fundamental, point is made by combining Equation (5.66) with the expression for μ_0 at low frequencies to give

$$f_0 = \frac{4}{3}\frac{\gamma I_s}{\mu_0 - 1} . \tag{5.67}$$

where $f_0 = \omega_0/2\pi$. Thus, f_0 is the effective 'limiting frequency' of operation and is reduced by high values of μ_0. The effect is shown very well by Figure 5.33. Taking f_0 as the frequency at which μ' had fallen to half its original value, Smit and Wijn obtained excellent agreement between the values of μ_0 measured from the figure and those calculated from Equation (5.67) [34].

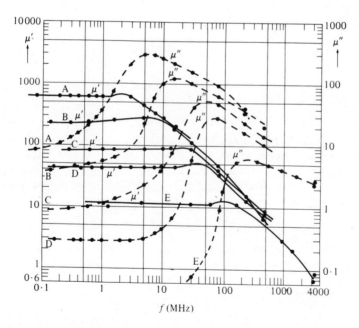

Figure 5.33. Permeability spectra for a range of commercial nickel–zinc ferrites, namely Philips Ferroxcube IV, grades A–E, showing peaks in the losses due to resonance in the anisotropy fields. The composition may be represented by $Ni_\delta Zn_{1-\delta}Fe_2O_4$ where $\delta = 0.36$, 0.5, 0.64, 0.8 and 1.0 for A to E respectively. Note the different scales for μ' and μ''. (Smit and Wijn [34]).

To a certain extent this limitation has been overcome by fabricating special hexagonal oxides (e.g. for 'Ferroxplanas') with easy planes of magnetization [35]. The ease of rotation of the magnetization within the basal plane gives a relatively high rotational permeability. Precession must involve rotation out of the easy plane; this is opposed by a high anisotropy and gives high resonant frequencies. Thus, the apparently basic restriction outlined above is circumvented by recourse to this particular type of anisotropy.

The topic of ferromagnetic resonance is taken further in Chapter 6 where a description is given of domain wall resonance and relaxation. These phenomena also affect the permeability spectra, such as those shown in Figure 5.33. Even static walls have some effect, since they alter the environment in which the magnetization rotates. Ferromagnetic resonance experiments as such are generally carried out in the presence of large static fields which cause saturation and eliminate the complications associated with domain walls. The spread of the losses, around the resonance frequency, or linewidth, is connected with loss mechanisms as already discussed and also with a number of structural factors. In a polycrystal the randomly oriented anisotropy fields give different contributions in the direction of the static field, and thus cause an effective spread in the total field, and in the values of the applied field for which the losses are substantial. The stray fields arising from pores have a similar effect.

The effect of structural factors on the resonant frequencies can be represented by an internal field H_i, such that

$$\omega_0 = \gamma(H_z + H_i) ; \qquad (5.68)$$

H_z is the applied field, assumed to be in the OZ direction. Kittel showed how the simple relation, $\omega_0 = \gamma H_z$, which applies to a spherical specimen with negligible crystal anisotropy, should be modified to account for the demagnetizing fields of ellipsoidal specimens with demagnetizing factors N_x, N_y, N_z, namely:

$$\omega_0 = \gamma\{[H_z + (N_x - N_z)I_s][H_z + (N_y - N_z)I_s]\}^{\frac{1}{2}} \qquad (5.69)$$

and this must be further modified, for a single crystal, to account for the anisotropy torques, giving

$$\omega_0 = \gamma\left\{\left[H_z + (N_x - N_z)I_s + \frac{2K_1}{I_s}\cos 4\theta\right]\right.$$
$$\left. \times \left[H_z + (N_y - N_z)I_s + \frac{K_1(3 + \cos 4\theta)}{I_s}\frac{}{2}\right]\right\}^{\frac{1}{2}} \qquad (5.70)$$

for fields applied at an angle θ to a [100] easy direction [36].

Thus, both demagnetizing effects and crystal anisotropy can shift the resonance frequency with respect to that for an isotropic sphere. For a thin isotropic sheet situated in the OXZ plane, as shown in Figure 5.34, $N_x = N_z = 0$ and $N_y = 4\pi$ so that

$$\omega_0 = \gamma[H_z(H_z + 4\pi I_s)]^{\frac{1}{2}} , \qquad (5.71)$$

while for a sheet normal to OZ, $N_x = N_y = 0$ and $H_z = 4\pi$, whence

$$\omega_0 = \gamma(H_z - 4\pi I_s) , \qquad (5.72)$$

which is just the internal field, and for a cylinder with its long axis parallel to OZ, $N_x = N_y = 2\pi$, $N_z = 0$ and

$$\omega_0 = \gamma(H_z + 2\pi I_s) . \qquad (5.73)$$

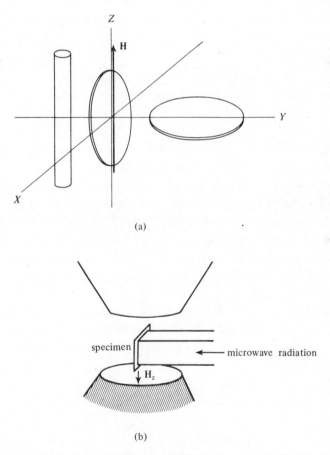

(a)

(b)

Figure 5.34. (a) Disposition of the plate-shaped or rod-shaped specimens with respect to the applied d.c. field in resonance experiments, the microwave radiation being normal to OZ. (b) Schematic diagram of apparatus for resonance experiments, showing the magnet pole pieces and microwave cavity.

If internal demagnetizing fields exist, due to porosity or gross defects, their variation in direction and magnitude will give a spread in the value of ω_0, or conversely will give a range of fields over which resonance can be obtained, and thus broaden the linewidth for absorption. Similarly, in a polycrystal which has finite anisotropy it is found by averaging formulae of the type given above, that there is a shift in the field for resonance given by [see Equation 5.68)]

$$H_i = -\frac{K_1}{2I_s} , \tag{5.74}$$

together with a contribution to the linewidth due to the random orientation of the anisotropy fields of

$$\Delta H = \frac{\sqrt{3}[H_z^2 - H_z(\frac{4}{3}\pi I_s) + \frac{5}{12}(\frac{4}{3}\pi I_s)^2]}{[(H_z - \frac{8}{3}\pi I_s)(H_z - \frac{4}{3}\pi I_s)^3]^{\frac{1}{2}}} \frac{H_A^2}{4\pi I_s} \tag{5.75}$$

for a spherical specimen, where H_A is the anisotropy field [37].

Schlömann [38] calculated the value of H_i due to a porosity p (the ratio of the volume of the pores to the total volume of material) as

$$H_i = \frac{4}{3}\pi I_s p ,$$

with a contribution to the linewidth of

$$\Delta H = 1 \cdot 5(4\pi I_s)p .$$

The most striking demonstrations of the effects of these structural factors arise from studies on yttrium iron garnet, since it is found that stoichiometric single crystals of this material can have linewidths well below 1 Oe (see Figure 5.35). Extreme values of porosity can give

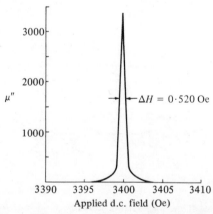

Figure 5.35. Ferromagnetic resonance absorption line for a polished single-crystal sphere of yttrium iron garnet, as usually observed at a fixed microwave frequency (9300 MHz) and a variable d.c. field. ΔH is defined as the width of the line at half its height (LeCraw *et al.* [40]).

linewidths of hundreds of oersteds, and even with $p \approx 1\%$, $\Delta H \approx 10$ Oe. Anisotropy broadening in polycrystalline YIG is approximately 10 Oe at room temperature, so both effects can be overwhelming in sintered polycrystals. In materials such as the nickel–zinc ferrites (for example Ferroxcube 4A) a substantial contribution to the linewidth is probably derived from the intrinsic damping associated with the ferrous ion content, but porosity effects are still considerable: linewidths of 40 to 60 Oe are observed for specimens with $p < 0.05\%$, while commercial specimens have $\Delta H \approx 200$ Oe [39].

It remains to consider why YIG should have such an exceptionally low intrinsic linewidth. First, it is possible to prepare the garnets with a very high degree of stoichiometry, the conditions of sintering or crystal growth being much less critical than for the spinel ferrites. Furthermore, the ordering of the ions in garnets is extremely regular. Thus, the presence of ferrous ions is readily avoided: ideally YIG contains only Y^{3+} and Fe^{3+} cations.

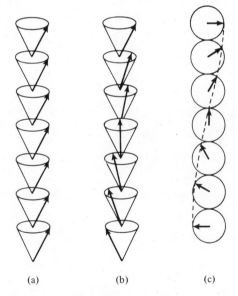

Figure 5.36. Schematic illustration of spins precessing with negative phase differences to give spin waves: (a) shows the coherent mode and (b) shows one half wavelength. (c) shows the spins in plan.

Now it must be remembered that in order to absorb energy from an r.f. field or from microwave radiation, this energy must be passed to the crystal lattice and appear as heat. The coupling between spins and the crystal lattice is indirect, via the spin–orbit coupling, but in YIG all the cations are in S-states with no orbital angular momentum, and thus the coupling is very low. It is hardly surprising to find that certain impurities can give a substantial broadening, and that Co^{2+} is outstanding in this respect as are the rare-earth ions with large unquenched angular momenta (such as holmium).

For any further discussion spin waves must be introduced. It has so far been assumed that the ionic spins precess in phase (the uniform mode), but they may precess with a regular phase difference throughout the crystal (Figure 5.36), or a phase difference may be propagated as a spin wave through the crystal. These spin waves are quantized; just as the vibrations of a crystal lattice may be described in terms of quanta of vibrational energy, or phonons, the spin waves may be represented as magnons with energies which depend on the wavelengths of the spin waves, or the wave vectors k. The uniform mode can be considered as the zero k spin wave. The higher order spin waves may have associated magnon energies which are degenerate with the phonon energies of the crystal lattice, and the transfer of energy is thus facilitated by any process which excites these rather than the uniform mode. Irregularities in the ordering of the ions can excite spin waves, again indicating the value of YIG. Spin waves may also be excited by surface pits, and the narrowest lines of all are obtained only with highly polished spheres of high purity single crystals. ΔH may then be as low as $0 \cdot 01$ Oe [40].

References

1. E. C. Stoner and E. P. Wohlfarth, *Phil. Trans. Roy. Soc.*, **A240**, 599 (1948).
2. R. Gans, *Ann. Physik*, **15**, 28 (1932).
3. L. Néel, *Compt. Rend.*, **228**, 604 (1949); *Ann. Geophys.*, **5**, 99 (1949).
4. W. F. Brown, *J. Appl. Phys.*, **30**, 130S (1959).
5. D. J. Craik, D. D. Eley and R. J. Mellar, *Trans. Faraday Soc.*, **65**, 1649 (1969).
6. W. F. Brown, *Phys. Rev.*, **105**, 1479 (1957); *J. Appl. Phys. Suppl.*, **30**, 62 (1959); E. H. Frei, S. Shtrikman and D. Treves, *Phys. Rev.*, **106**, 446 (1957); A. Aharoni, *J. Appl. Phys. Suppl.*, **30**, 70 (1959).
7. W. F. Brown, *"Magnetostatic Principles in Ferromagnetism"* (North Holland, Amsterdam), 1962; *"Micromagnetics"* (Interscience, New York), 1963.
8. I. S. Jacobs and C. P. Bean, *Phys. Rev.*, **100**, 1060 (1955).
9. E. P. Wohlfarth, *"Magnetism"*, Volume III, Eds. G. Rado and H. Suhl (Academic Press, New York), 1963.
10. F. E. Luborsky and C. R. Morelock, *J. Appl. Phys.*, **35**, 2055 (1964).
11. F. E. Luborsky, *J. Appl. Phys. Suppl.*, **32**, 171 (1961).
12. K. J. de Vos, *"The Relationship between Microstructure and Magnetic Properties of Alnico Alloys"*, Thesis, Eindhoven, 1966.
13. J. E. Gould, *Cobalt*, **23**, 1 (1964).
14. J. Harrison and W. Wright, *Cobalt*, **35**, 63 (1967).
15. J. E. Gould, *Proc. Inst. Elec. Engrs. (London)*, **106A**, 493 (1959); and see also reference 13.
16. F. Anselin, *Cobalt*, **4**, 3 (1959).
17. W. H. Meiklejohn and C. P. Bean, *Phys. Rev.*, **102**, 1413 (1956); **105**, 904 (1957).

18. C. Guillaud, *"Ferromagnetism of Manganese Binary Alloys"*, Thesis, Strasbourg, 1943.
19. C. D. Mee and J. C. Jeschke, *J. Appl. Phys.*, **34**, 1271 (1963).
20. G. S. Kritjenburg, Proc. 1st European Conf. Permanent Magnets, Vienna, 1965.
21. K. Strnat, G. Hoffer, J. Olson, W. Ostertag and J. J. Becker, *J. Appl. Phys.*, **38**, 1001 (1967).
22. D. J. Craik and D. A. McIntyre, *Proc. Roy. Soc.*, **A302**, 99 (1967).
23. D. J. Craik and D. A. McIntyre, *Physics Letters*, **21**, 288 (1966).
24. S. Shtrikman and D. Treves, *J. Appl. Phys. Suppl.*, **31**, 72 (1960).
25. H. Zijlstra, *Z. Angew. Phys.*, **21**, 1, 6 (1966).
26. A. Aharoni, *J. Appl. Phys.*, **33**, 1324 (1962).
27. E. V. Shtol'ts, *Sov. Phys.–Solid State (English Transl.)*, **8**, 2738 (1967).
28. L. Landau and F. Lifshitz, *Physik. Z. Sowjet.*, **8**, 153 (1935).
29. T. L. Gilbert, *Phys. Rev.*, **100**, 1243 (1955).
30. R. Kikuchi, *J. Appl. Phys.*, **27**, 1352 (1956).
31. J. K. Galt, *Phys. Rev.*, **85**, 664 (1952); *Proc. Inst. Elec. Engrs. (London)*, **104B**, 189 (1957).
32. J. L. Snoek, *"New Developments in Ferromagnetic Materials"* (Elsevier, New York), 1947.
33. H. P. J. Wijn and H. Van der Heide, *Rev. Mod. Phys.*, **25**, 98 (1953).
34. J. Smit and H. P. J. Wijn, *Advan. Electron. Electron Phys.*, **6**, 69 (1954).
35. G. H. Jonker, H. P. J. Wijn and P. B. Braun, *Philips Tech. Rev.*, **18**, 145 (1956); *Proc. Inst. Elec. Engrs. (London)*, **104B**, 249 (1957); A. L. Stuijts and H. P. J. Wijn, *Philips Tech. Rev.*, **19**, 209 (1957/1958).
36. C. Kittel, *Phys. Rev.*, **71**, 270 (1947); **73**, 155 (1948).
37. K. J. Standley and K. W. H. Stevens, *Proc. Phys. Soc. (London)*, **69B**, 993 (1956); E. Schlömann, *Phys. Chem. Solids*, **6**, 242 (1958); E. Schlömann, J. J. Green and U. Milano, *J. Appl. Phys.,Suppl.*, **31**, 386 (1960).
38. E. Schlömann, Raytheon Tech. Rept. R-15, 1st September 1956.
39. A. J. Stuijts, J. Verweel, and H. P. Peloschek, *I.E.E.E. Trans. Commun. Electron.*, **83**, 726 (1964).
40. R. C. LeCraw, E. G. Spencer and C. S. Porter, *Phys. Rev.*, **110**, 1311 (1958); E. G. Spencer and R. C. LeCraw, *Proc. Inst. Elec. Engrs. (London)*, **109B**, 66 (1961).

6

Domain Structure

6.1 INTRODUCTION

It is a matter of common observation that, so long as they are not very small, ferromagnetic or ferrimagnetic crystals may be readily demagnetized and that the demagnetized state corresponds to subdivision into domains. Typical structures, both of which obey the requirement that the total resolved magnetization in any direction is zero, are shown in Figures 6.1a and b. The different types of structure correspond principally to different types of anisotropy. When we say that a certain structure is typical of a certain specimen, we imply that it has a low energy, the appropriate energy terms being those for the domain walls,

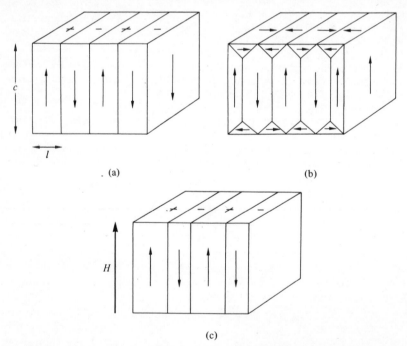

Figure 6.1. Representative domain structures, greatly simplified in comparison with those in most real crystals: (a) Uniaxial crystal, with 180° domains only. (b) Crystal with cubic anisotropy (positive K_1) showing flux closure via the 90° domains at the surface. (c) Partially magnetized uniaxial crystal with the 180° walls displaced by an applied field.

the anisotropy energy, the magnetostrictive energy, and the demagnetizing energy, while exchange energy only appears indirectly through its contribution to the energy of the walls.

Thus, Figure 6.1a is a typical uniaxial structure of $180°$ domains. There is obviously a fair demagnetizing energy associated with the surface poles, which does not arise in Figure 6.1b. However, if the structure of Figure 6.1b were considered to apply to a uniaxial material, a large value of the crystal anisotropy energy would be involved unless K_1 and K_2 were very low, which is not generally the case for uniaxial materials. Hence, Figure 6.1b is typical for cubic crystals with positive K_1, the only energy involved in the closure structures being magnetostrictive in origin.

The demagnetizing energy (Figure 6.1a) or magnetostrictive energy (Figure 6.1b) could be reduced indefinitely by packing in more and more $180°$ walls. This procedure, however, would raise the wall energy, and the equilibrium spacing which is expected (although not always achieved in practice, depending on the magnetic or thermal treatment) is that for which the sum of the appropriate energy terms is a minimum.

As well as providing a necessary explanation for the possibility of demagnetizing crystals without destroying the spontaneous magnetization, the presence of domains also explains the very high susceptibilities which may be achieved, even when the crystal anisotropy is substantial. If a field is applied parallel to the easy axis in Figure 6.1a, the energy of the magnetization in the field is $\int \mathbf{H} \cdot \mathbf{I}_s \, dv$, which is zero if the walls remain arbitrarily fixed in position. If each wall moves through a distance x, as shown in Figure 6.1c, then

$$E_H = 2HI_s xA \; ,$$

assuming that the total area of the walls is A. Thus, there is an effective pressure (force per unit area) on the walls, given by

$$-\frac{1}{A}\frac{\partial E_H}{\partial x} = 2HI_s \; . \tag{6.1}$$

If the energy of the domain walls is a function of their position, as indicated for one wall in Figure 6.2, then the wall will initially be in a 'trough', at x_1, and will be forced 'uphill' until a point of inflection is reached, at x_2, by an applied field. Beyond this displacement the wall will move freely in the same field to a point, x_3, at which the value of $\partial E/\partial x$ is the same, in a so-called *Barkhausen jump*. The variation of the energy with respect to the position of the wall clearly represents an impediment to its motion, or a restoring force which opposes the pressure due to the applied field. In a perfect crystal there should be no such energy variation, and consequently the susceptibility should be unlimited in the absence of external demagnetizing effects. In a real crystal imperfections will give rise to a finite coercivity and also limit the susceptibility.

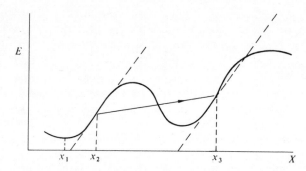

Figure 6.2. Schematic variation of the energy of a domain wall with respect to its displacement. E includes the demagnetizing energy of inclusions intersected by the wall etc. x_1 is a minimum and x_2 a point of inflection. The slope at x_3 is equal to that at x_2.

6.2 GENERAL FEATURES OF MAGNETIZATION BY WALL MOTION

Firstly, it is supposed that the energy variation associated with the position of the wall, $E(x)$, is regularly periodic. This will mean that whatever defects exist they are uniformly distributed throughout the specimen, which is taken initially to be a single crystal with an easy axis parallel to the field direction, or a polycrystal in which all the grains are crystallographically aligned to give the same effect. The amplitude of $E(x)$ will correspond to a certain critical field H_0, given by

$$2H_0I_s = \left[\frac{\partial E(x)}{\partial x}\right]_{max} , \qquad (6.2)$$

for which this and all succeeding barriers will be surmounted. Thus, the walls move throughout the crystal in this same field to give complete saturation as shown in Figure 6.3. H_0 also represents the coercivity since, assuming negligible nucleation fields, the walls move once more throughout the specimen in fields of $-H_0$.

It is next supposed that the specimen has a finite demagnetizing factor, so that, as magnetization proceeds, the demagnetizing energy, E_d, increases. While the foregoing would refer to a ring-shaped specimen (wound from grain-oriented alloy tape for example) the demagnetizing energy would arise in a rod of finite length or in a ring containing a gap. E_d is a smoothly varying function of the position of the walls, which means that, in contrast to $E(x)$, it is non-periodic, and is thus equivalent to a restoring force which affects the wall motion throughout magnetization and it can be shown to give an anhysteretic magnetization curve of the type indicated by the broken line in Figure 6.3. (An anhysteretic curve is one for which the coercivity is neglected or, in the experimental case, suppressed by the simultaneous application of alternating and static fields.) When the two forms of energy variation are considered together, it must be appreciated that the walls do not commence to move until the field H_0 is reached, and a little consideration shows how the sheared loop indicated in Figure 6.3 is built up. As the fields are reduced after saturation, if $H_d > H_0$, the walls may commence to move in positive fields but the coercivity will still be H_0 since as demagnetization is approached $H_d \to 0$.

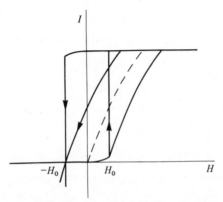

Figure 6.3. Schematic magnetization loop and initial magnetization curve for a uniaxial material with uniform impediments to domain wall motion and zero demagnetizing factor (upright loop). The broken line has the general shape of anhysteretic magnetization curves calculated for specimens with finite demagnetizing factors, and the sheared loop is built up from this.

A great simplification arises if the specimen is ellipsoidal and is large compared with the domain spacing. In this case the demagnetizing fields are uniform, $H_d = NI$, and the anhysteretic initial and differential susceptibilities are given by

$$\chi = \frac{I}{NI} = \frac{1}{N} \ . \tag{6.3}$$

If the effect of the coercivity is included, the maximum susceptibility is

$$\chi_m = \frac{I_s}{H_0 + NI_s} \ , \tag{6.4}$$

since the magnetization curve is a straight line up to saturation in applied fields given by $H_a - NI_s = H_0$.

Somewhat more elaborate analysis is necessary for a specimen such as randomly oriented iron. The effective pressure due to the field becomes $2HI_s \cos\theta$, where θ is the angle between the field and the nearest easy direction, and this will give a range of fields necessary to initiate wall motion even with uniform energy barriers. Thus, the steeply rising part of the magnetization curve is sheared even when demagnetizing energy is neglected. However, demagnetizing energy cannot in fact be neglected because, even if the overall demagnetizing factor is zero, considerable demagnetizing energy arises from the total normal components of the magnetization at the grain boundaries and external surfaces. These latter effects appear to predominate in practice, because it is found experimentally, for silicon iron for example, that hysteresis loops can be constructed by shifting the measured anhysteretic curves through $\pm H_c$, as shown in Figure 6.4.

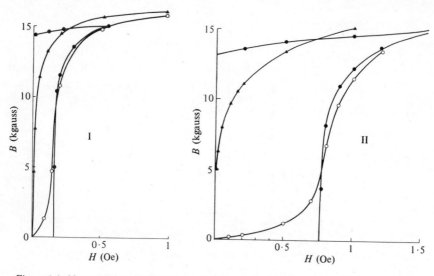

Figure 6.4. Magnetization loops (●), magnetization curves (○) and anhysteretic magnetization curves (▲) for two samples of grain-oriented silicon–iron with identical composition, similar degrees of orientation, but differing grain diameter (0·56 mm for specimen I and 0·025 mm for specimen II). In both cases the side of the hysteresis loop is very nearly parallel to the anhysteretic curve; on the assumption that the slope and curvature of the anhysteretic curve is due to the build-up of magnetostatic energy and cannot have any relation to the spread of impediments to domain wall motion, the shape of the sides of the hysteresis loop can be interpreted similarly as being due to magnetostatic effects. (Measurements kindly made by A.E.I. Laboratories on specimens supplied by Martin Littmann, Armco Co.).

It is also to be noted that fields of the order of 10 Oe can cause little rotation in most materials but, at the most, can cause saturation of each crystallite along an easy direction which is nearest to the field direction. The level of magnetization for random orientation is then given by the same calculation as that which gives the remanence of single-domain particles. Clearly, polycrystals with cubic crystal structure can always be considered to have a certain degree of natural orientation, since there is always an easy direction within 55° of any applied field direction. For randomly oriented uniaxial polycrystals half the saturation magnetization must be achieved by rotation, and any attempts at analysis are very difficult.

For a great number of materials, however, it is legitimate to deal with the coercivity and the effects of demagnetizing fields separately, on the assumption that the two can be superimposed to give the hysteresis loops, but before this treatment the domain walls must be described more fully.

6.3 STRUCTURE AND PROPERTIES OF DOMAIN WALLS

If the magnetization is arbitrarily fixed in antiparallel directions, as shown at the extremes of the block in Figure 6.5a, then the rotation of the magnetization in between must involve both exchange energy (since the spins are not parallel to each other) and anisotropy energy (since the spins diverge from the easy direction). For unit area of wall, with the axis OZ normal to the wall and the origin in the centre of the wall, the exchange energy is given by

$$W_e = A \int_{-\infty}^{\infty} \left(\frac{\partial \theta}{\partial z} \right)^2 dz \ , \tag{6.5}$$

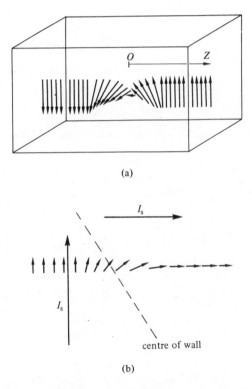

(a)

(b)

Figure 6.5. (a) The gradual rotation of the spins or atomic magnetic moment vectors forming the transition between antiparallel magnetization directions, that is a 180° domain wall or Bloch wall. (b) The structure of one type of 90° domain wall.

where θ is the azimuthal orientation of the spins, and changes from $\pi/2$ to $-\pi/2$. The anisotropy energy is

$$W_a = \int_{-\infty}^{\infty} g(\theta)\,dz \ , \tag{6.6}$$

where $g(\theta) = K\cos^2\theta$ for uniaxial anisotropy. Landau and Lifshitz [1] first showed how the actual (minimum energy) configuration of the spins could be found by applying Euler's method to the sum of these two terms, giving as a result a wall with a specific energy and a certain width within which most of the rotation occurs. Empirically, the anisotropy energy tends to narrow the wall, while the exchange energy tends to spread it out. The Euler equation is

$$\frac{d^2\theta}{dz^2} = \frac{K}{2A}\sin 2\theta \ ,$$

with solutions $\theta = 0$, $x\pi$, and $\sin\theta = \dfrac{\pm 1}{\cosh(K/A)^{1/2}(z+z')}$, where z' is a constant of integration. The first two solutions give zero energy and are neglected. The other two both give an energy

$$\gamma = 4(AK)^{1/2} \ \text{erg cm}^{-2} \ , \tag{6.7}$$

and the corresponding length over which the rotation from $\theta = 0$ to $\theta = 180°$ occurs, that is the $180°$ domain wall width, is

$$\delta = \pi(A/K)^{1/2} \ \text{cm} \ . \tag{6.8}$$

For a $180°$ wall in a cubic material, with the magnetization rotating in a (100) plane,

$$\gamma = 2(AK)^{1/2} \ \text{erg cm}^{-2} \ , \tag{6.9}$$

while for a $90°$ wall (Figure 6.5b)

$$\gamma = (AK)^{1/2} \ \text{erg cm}^{-2} \ . \tag{6.10}$$

Further results are given by Lilley [2].

Taking A equal to 10^{-6} erg cm, which applies quite closely to most materials [3], one obtains widely varying values for the wall parameters of practical materials, corresponding to the wide variation of crystal anisotropy. In the case of highly anisotropic materials $\delta \doteq 10^{-6}$ cm and $\gamma \doteq 1$ erg cm^{-2}.

Where a $180°$ wall intersects a surface parallel to the easy axis, the strip of pole density formed must give rise to some demagnetizing energy. Since the wall width is much smaller than the wall spacing in most specimens, this is negligible. In very thin specimens, such as evaporated films, demagnetizing energy may predominate and Néel walls, with the manner of rotation illustrated by Figure 6.6, have the lower energy [4].

Figure 6.6. The structure of a 180° Néel wall as found in very thin magnetic films. The rotation occurs in the plane of the film, since this creates less demagnetizing energy than would the strips of free pole at the surface intersections of a Bloch wall.

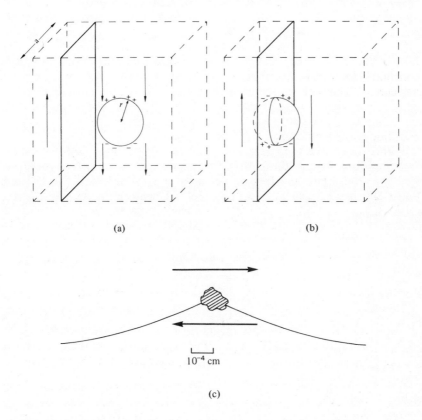

(a)   (b)

(c)

Figure 6.7. When a domain wall lies near to a spherical pore or non-magnetic inclusion (a) the energy associated with the arbitrary cube of material is $a^2\gamma + \frac{1}{2}(\frac{4}{3}\pi)(\frac{4}{3}\pi r^3)I_s^2$, but when the wall bisects the pore (b) this is reduced to $(a^2 - \pi r^2)\gamma + \frac{1}{4}(\frac{4}{3}\pi)(\frac{4}{3}\pi r^3)I_s^2$. (c) The actual extent of wall deformation produced by an inclusion in uniaxial magnetoplumbite, when the wall is driven past the inclusion by an applied field (tracing of powder pattern).

6.4   IMPEDIMENTS TO WALL MOTION: WALL COERCIVITY

The most obvious type of gross defect is a hole or pore, of the type which does actually exist in many ceramic materials or, equivalently, a non-magnetic inclusion. Typical inclusions of real importance are nitrides and carbides in iron and its alloys. If a domain wall comes into promixity with a spherical hole it will tend to assume a position in which it bisects the hole (Figure 6.7) for two reasons.

The first and most apparent reason is that the maximum proportion of the total wall area is then effectively eliminated, and the effect is similar to that of surface tension in a liquid film. The form of $E(x)$ due to this follows from simple geometry, if it is assumed that the wall remains planar. The second and more generally important effect follows from the form of the demagnetizing energy which is associated with the hole. This is identical to that of a magnetized sphere, since the pole densities are equivalent in the two cases, and in each case the demagnetizing energy is reduced to the greatest extent when the wall is central. Néel [5] showed that for holes or inclusions in iron of 10^{-4} cm diameter, the effect of the demagnetizing energy is a hundred times greater than that of the changes in wall energy, but for materials with low saturation magnetization and high wall energy (that is with high anisotropy) the wall energy effect is not negligible.

Non-uniform internal stresses (σ_i) impede wall motion, since substantial deviations in the direction of the magnetization (which occur if $\lambda_s \sigma_i \doteq K_1$) also give rise to pole densities at the borders of the stressed region. Extensive calculations by Néel [5] dealt both with inclusions occupying a fractional volume v_i and stresses with large values in a fractional volume v_σ of the material, and gave the results for iron

$$H_c = 2 \cdot 1 v_\sigma + 360 v_i ,$$

and for nickel

$$H_c = 330 v_\sigma + 97 v_i .$$

However, it does not seem advisable to pursue the quantitative aspect of coercivity very far; in soft magnetic materials low coercivities and high permeabilities are required and the major interest in inclusions and internal stresses lies in the means of their elimination or in measures to make them less effective.

Of major importance in this latter respect is the observation by Dijkstra and Wert [6] and by Tebble [7] that the impeding effect of imperfections is greatest when their dimensions are comparable to the domain wall thickness. Inclusions much smaller than δ should have little effect. (Single dislocations can apparently be included under this heading, but dense clusters of dislocations can constitute a considerable wall impedance.)

6.5 SOFT MAGNETIC MATERIALS

Soft magnetic materials are basically required to have high initial or maximum permeabilities or susceptibilities and low losses; that is to say, the permeabilities should remain high over a considerable range of frequencies. The relation of initial permeabilities to imperfections is the converse of that of coercivity; generally a low coercivity can be equated to a high initial permeability. For a high maximum permeability it is also necessary that nearly complete saturation should be achieved in fields little higher than H_c. Thus, the magnetization curve should be steep, and according to section 6.2, this will be the case if extensive demagnetizing energies and the necessity for rotation are avoided: both of these considerations are met by producing materials with a grain-oriented structure.

Losses are reduced in two ways. Eddy-current losses in alloys are principally limited by using the alloys in the form of laminations or thin strips which are insulated from each other. For 50 Hz transformers the laminations may be quite substantial, about 1 mm thick, since this is found to give higher maximum permeabilities and is more economical, while for kilohertz frequencies the thickness must be approximately 10^{-2} mm as in high quality tape-wound cores. For very high frequencies, up to megahertz values, low conductivity ferrites are more commonly utilized, for which no lamination is necessary, and at microwave frequencies the use of near-insulators is essential.

The important features of the development of soft magnetic materials can be illustrated by reference to just four particular studies.

6.5.1 The effect of purifying metals

Figure 6.8 really requires little comment. The purification is effected by heating in hydrogen, which removes the impurities as they diffuse to the surface and thus precludes the reformation of inclusions of Fe_3C, Fe_3N etc. It may be noted that such treatment is costly, and that iron of high purity is not in fact a material of wide technical usefulness, in spite of the properties which may be achieved: for example $\mu_0 = 4000$, $\mu_m = 180000$ and the hysteresis loss is 190 erg cm^{-3} per cycle [8].

6.5.2 The effect of reducing the anisotropies in nickel–iron alloys and manganese–zinc ferrite

With appropriate heat treatment the crystal anisotropy of nickel–iron alloys passes through zero near to the composition with 75% Ni. The magnetostriction is also zero near to this composition, namely at 82% Ni, and it is hardly surprising that a peak in the permeability occurs at about 78% Ni (or 'Permalloy') as shown in Figure 6.9. Clearly, it is the total anisotropy and not just the crystal anisotropy which is of importance.

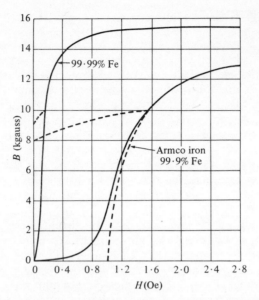

Figure 6.8. Loops representing the great advance made in the 1930's, and illustrating the remarkable sensitivity of the coercivity, and thus the maximum permeability, to small amounts of impurity (presumably largely carbon).

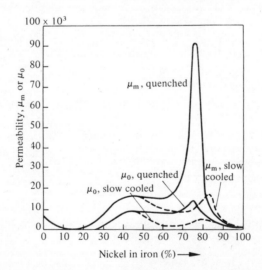

Figure 6.9. Peaks in the permeability of nickel–iron alloys corresponding to the compositions with minimum total anisotropy [R. M. Bozorth].

The significance of the heat treatment arises from the tendency to produce either short-range or long-range ordering of the nickel and iron atoms, which gives rise to considerable anisotropies, if the alloys are maintained at temperatures at which substantial atomic diffusion can occur. The double heat treatment referred to in Figure 6.9 consisted of annealing to remove stresses followed by rapid cooling through the range of temperatures in which ordering can occur.

It might appear that the high permeabilities correspond to rotation against negligible anisotropies (section 5.3). However, there is indirect evidence for wall motion and the author's opinion is that the real effect of the low anisotropy is to make the walls so wide, and of such low energy, that any inclusions present have little effect on their motion: possibly this will be confirmed by direct domain studies in the future.

The inclusion of other elements, for example molybdenum to give 'Supermalloy' (79Ni 16Fe 5Mo), can simplify the heat treatment by modifying the ordering, and also lead to the coincidence of the zero anisotropy and zero magnetostriction compositions. Maximum permeabilities of 10^6 can then be achieved.

The crystal anisotropy of manganese–zinc ferrites also passes through zero at temperatures which depend on the precise composition [9]. The low anisotropies are associated with the presence of ferrous ions, which give rise to substantial high frequency losses (that is corresponding to low resistivities for ferrites, around 10^2 ohm cm; cp 10^7 or more for Ni ferrite). Thus, these materials fulfil an intermediate role between the alloys, which are generally preferable at low frequencies because of their high saturation, and the high-frequency ferrites, which have very low conductivity and low losses.

6.5.3 Grain-oriented silicon iron

The inclusion of silicon in iron confers several technical advantages. The resistivity is raised with respect to that of pure iron, and there is a reduction in eddy-current losses. The anisotropy is reduced and the formation of inclusions is partially suppressed to give higher permeabilities. Furthermore, the suppression of the phase changes which occur in pure iron makes it possible to produce grain-oriented material. The concentration of silicon is usually around 3% in technical materials, but there is also interest in steels containing much higher concentrations [10] although these are brittle and very difficult to work.

Grain orientation is produced by a carefully controlled routine of cold rolling and annealing. The degree of cold reduction, the presence of impurities, and the nature of the annealing atmosphere are among the important parameters. The two basic principles involved are that plastic deformation can itself give rise to a certain texture, and that this can be reinforced or modified by annealing due to the dependence of the surface energy on the crystallographic orientation of the surface.

The two types of texture, namely single orientation or [100](110), and cube texture or [100](100), are illustrated by Figure 6.10: note that the former contains only one easy direction parallel to the surface of the sheet. Since the 'cross directions' are unlikely to be occupied on magnetostatic grounds, such materials are highly anisotropic and only have high permeabilities when fields are applied parallel to the common [100] direction. Fortunately, this is also the rolling direction and, in consequence, that which is favourable for the production of long strips.

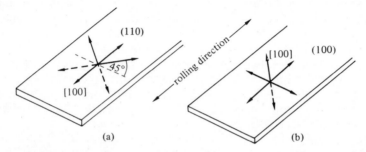

Figure 6.10. (a) In singly-oriented or 'Goss-textured' silicon–iron, crystallites have surfaces which are predominantly {110} and thus contain a ⟨100⟩ easy direction, whereas other easy directions are at 45° to the surface and unlikely to be substantially occupied on magnetostatic grounds. (b) In cube-textured silicon–iron the crystallite surfaces are close to {100} planes and thus there are two easy axes parallel to the surface of the sheet.

The singly oriented material is in very common use, its applications ranging from heavy duty transformers to high quality tape cores for magnetic amplifiers [11]. A certain reticence seems to apply to the production and utilization of cube-textured material. Assuming that it could be produced in large quantities, its real advantages would only be realized by stamping out continuous frames (with no air gaps at all) but subsequent production difficulties would then be encountered, for example in the application of the windings. Also, the eddy-current losses generally seem to be higher in cube-textured alloys than in singly oriented material.

The ordinary and anhysteretic magnetization loops for experimental specimens of oriented silicon iron prepared by Martin Littmann of the Armco Company [12], have already been given in Figure 6.4. By comparing the two curves it is possible to decide the extent of the improvement which should accrue either from decreasing the coercivity or from producing a higher degree of alignment. It must be remembered that the utilization of silicon iron in transformers calls for high maximum permeabilities and these (as contrasted with initial permeabilities) are extremely sensitive to the degree of alignment. The values which can be obtained represent a great technical achievement, but if it is assumed that the properties of a perfectly grain-oriented strip should be identical with those of a single crystal, then there is still much room for improvement: single crystals may have maximum permeabilities of 10^6 [13].

6.5.4 Porosity in ferrites

Porosity is not generally a factor of importance in metals, being appreciable only in specimens produced by powder metallurgy. Ferrites or other magnetic oxides are usually prepared by heating a mixture of the appropriate oxides to produce a diffusional solid state reaction, which is accelerated by prior milling to reduce the particle size and cause intimate mixing. The sintered or pre-fired material is milled a second time, pressed to the desired shape and finally sintered at temperatures up to 1400°C in atmospheres with controlled oxidizing or reducing characteristics. Ceramic materials prepared by these techniques always have some porosity, and in common commercial components this may be several per cent of the volume.

Needless to say, this is usually a disadvantage (but compare with section 6.8). The effect may only be fully appreciated when dense specimens can also be prepared by special methods and used for comparison. Great technical advances have recently been made by the

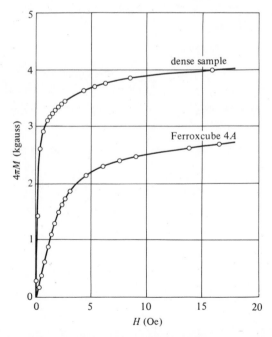

Figure 6.11. The effect of porosity on the anhysteretic magnetization curves of nickel–zinc ferrites, comparing a standard commercial product (Ferroxcube) with experimental material of very high density. There is also a substantial effect on the coercivity (Stuijts *et al.* [14]).

Philips laboratory [14], and these are exemplified by the anhysteretic loops in Figure 6.11 and the corresponding micrographs, Figure 6.12, for commercial and very dense nickel–zinc ferrites. The precise method of preparation was not specified, but a major problem was recognized as discontinuous grain growth. After cold pressing there must be many pores in between the grains. If these remain at the grain boundaries, there is a possibility of their being eliminated during the growth of the grains, but unless special attention is given to the details of preparation there is a tendency for numbers of grains to fuse discontinuously and to form a giant grain within which the pores are inextricably trapped. Factors of importance are purity and uniformity of grain size of the raw materials, and uniformity of packing.

In his current work de Lau [15] makes much use of hot pressing in a heated alumina die to produce specimens of remarkable ceramic quality.

A somewhat different development stems from the realization that, for high initial permeabilities at least, the extent of the porosity is less

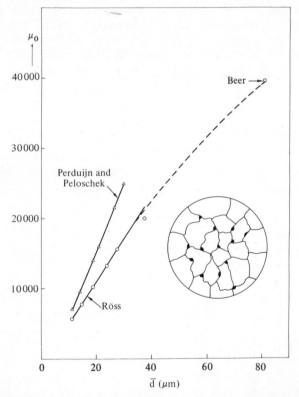

Figure 6.12. The variation of initial permeability with grain size for manganese–zinc ferrites having the microstructure represented by the sketch, that is with no intra-granular porosity. (Perduijn and Peloschek [16]).

important than its distribution. More specifically, pores within the grains are very deleterious, but even large pores at the grain boundaries have surprisingly little effect. Figure 6.12 shows a remarkably linear increase of initial permeability with grain size up to values of μ_0 which would have been considered outside the realms of possibility for ferrites some few years ago [16, 17, 18]. Such results apply only to specimens with the micro-structure shown by Figure 6.13 (between pp.212–213).

This latter information stresses a shortcoming of the discussion so far. Initial permeabilities have been considered to be limited only by the impeding effect of localized imperfections on the domain walls. When the imperfections are largely suppressed other factors become important, particularly the magnetostatic energies arising from the grain boundaries. This, and similar problems, require a detailed quantitative analysis of domain structures.

6.6   THE ANALYSIS OF DOMAIN STRUCTURES

If discussion is confined to $180°$ domains, no changes in anisotropy energy or magnetostrictive energy are involved. The only variable energy terms are those for the demagnetizing or magnetostatic energy, W_m, the energy of interaction with the magnetic field, W_H, and the domain wall energy, W_γ. Since W_H and W_γ are usually simple functions of the geometry, the analysis of domain structures and the derivation of anhysteretic magnetization curves consists mainly of problems in magnetostatics. (Magnetization curves follow directly from the calculated domain structures in the presence of applied fields.) The basic principle of calculation is the assumption that equilibrium is attained when the domain spacing, l, or wall position is such that

$$\frac{\partial}{\partial l}(W_m + W_\gamma) = 0 \ .$$

Initially it will be assumed that domain walls always remain planar, to give a structure of the type drawn in Figure 6.1a, and two problems will be investigated: (1) what is the domain spacing in a sheet of material of infinite lateral extent and depth c measured parallel to I_s, and (2) what is the equilibrium number of domains in a rectangular block of finite dimensions?

6.6.1 Infinite sheet

Fourier methods are applicable in this case, and it is suggested that reference be made to a standard work on Fourier analysis. With the coordinates as shown in Figure 6.14 it is assumed that there is no variation along OY or OZ and the charge density is periodic along OX

with period $2l$ (see also Figure 6.15). If the potential is ϕ, the energy per unit area is

$$W = \frac{1}{2l}\int_{-l}^{+l} \sigma(x)\phi(x,\tfrac{1}{2}c)\,dx \quad ; \tag{6.11}$$

σ can be represented by the Fourier expansion

$$\frac{\sigma}{I_s} = \tfrac{1}{2}a_0 + \sum_{1}^{\infty} a_n \cos\frac{n\pi x}{l} \quad , \tag{6.12}$$

which is standard for an even function, symmetrical about $x = 0$.

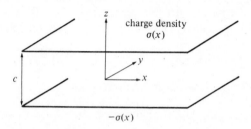

Figure 6.14. Position of coordinates for Fourier analysis, on the assumption that the charge density is periodic along OX, with period $2l$.

Next it is necessary to find ϕ as an appropriate solution of Laplace's equation (where in this case $\partial\phi/\partial y$ is equal to 0);

$$\frac{\partial^2\phi}{\partial x^2} + \frac{\partial^2\phi}{\partial z^2} = 0 \quad .$$

Standard trial solutions, for which $\phi \to 0$ as $z \to \pm\infty$ (since the net charge is zero and must give no effect at large distances) are:

$$\phi_1 = b_0 + \sum_{1}^{\infty} b_n \cos\frac{n\pi x}{l}\exp\frac{-n\pi z}{l} \qquad z > \tfrac{1}{2}c$$

$$\phi_2 = c_0 z + \sum_{1}^{\infty} c_n \cos\frac{n\pi x}{l}\sinh\frac{n\pi z}{l} \qquad \tfrac{1}{2}c > z > -\tfrac{1}{2}c$$

$$\phi_3 = d_0 + \sum_{1}^{\infty} d_n \cos\frac{n\pi x}{l}\exp\frac{n\pi z}{l} \qquad z < -\tfrac{1}{2}c \quad .$$

The boundary conditions are

$$\phi_1 = \phi_2 \qquad \text{at } z = \tfrac{1}{2}c \; ,$$

$$\left(\frac{\partial\phi_1}{\partial z}\right)_{z=\frac{1}{2}c} - \left(\frac{\partial\phi_2}{\partial z}\right)_{z=\frac{1}{2}c} = -4\pi\sigma \; ,$$

[1] Laplace's equation applies because $\mathbf{H} = -\mathrm{grad}\,\phi$ and $\mathrm{div}H = 0$, while $\nabla^2 = \mathrm{divgrad}$ (see Chapter 1).

and these boundary conditions give the coefficients

$$b_0 = \pi a_0 c \qquad\qquad b_n = 4\frac{l}{n} a_n \sinh\frac{\pi n c}{2l}$$

$$c_0 = 2\pi a_0 \qquad\qquad c_n = 4\frac{l}{n} a_n \exp\frac{-\pi n c}{2l}$$

$$d_0 = -\pi a_0 c_0 \qquad\quad d_n = -4\frac{l}{n} a_n \sinh\frac{\pi n c}{2l} \quad (= -b_n).$$

The magnetostatic energy per unit area is then given by

$$W = \frac{1}{2l}\int_{-l}^{+l} \sigma(x, \tfrac{1}{2}c)\phi(x, \tfrac{1}{2}c)\,dx$$

$$= \frac{1}{2l}\int_{-l}^{+l}\left(\tfrac{1}{2}a_0 + \sum a_n \cos\frac{n\pi x}{l}\right)$$

$$\times\left(\pi a_0 c + 2l\sum\frac{a_n}{n}\sinh\frac{\pi n c}{2l}\exp\frac{-\pi n c}{2l}\cos\frac{n\pi x}{l}\right)dx ,$$

and on carrying out the integration one obtains[2]

$$W = \tfrac{1}{2}\pi a_0^2 c + \sum_1^\infty a_n^2\frac{l}{n}\left(1 - \exp\frac{-\pi n c}{l}\right). \qquad (6.13)$$

If the charge density is represented by the odd function

$$\sigma = I_s\sum_1^\infty\left(a_n\cos\frac{n\pi x}{l} + b_n\sin\frac{n\pi x}{l}\right), \qquad (6.14)$$

the magnetostatic energy per unit area is

$$W = I_s^2 l\sum_1^\infty\frac{a_n^2 + b_n^2}{n}\left(1 - \exp\frac{-n\pi c}{l}\right). \qquad (6.15)$$

Expressions appropriate to particular domain structures can now be derived. [Note that Equation (6.15) applies to any periodic distribution and not to any specific structure.)

For parallel-sided slabs (as in Figure 6.15 with $l_1 = l_2$):

$$\sigma(x) = \begin{cases} -I_s & (-l, -\tfrac{1}{2}l) \\ +I_s & (-\tfrac{1}{2}l, +\tfrac{1}{2}l) , \\ -I_s & (+\tfrac{1}{2}l, +l) \end{cases} \qquad (6.16)$$

$a_0 = 0$ since $\sigma(x)$ averages to zero over one cycle, and

$$a_n = \frac{1}{l}\int_{-l}^{+l}\sigma(x)\cos\frac{n\pi x}{l}\,dx = \begin{cases} 0 & n \text{ even} \\ 4\pi I_s/n\pi & n = 4\kappa - 3 \\ -4\pi I_s/n\pi & n = 4\kappa - 1 \end{cases}\kappa = 1, 2, 3$$

[2] When the bracketed expressions are multiplied out prior to integration, terms containing $a_0 a_n$ are omitted (see, for example, N.K.Bary, 1964, *A Treatise on Trigonometric Series* (Pergamon Press, Oxford), p.65 *et seq.*, p.225 *et seq.*

and with Equation (6.13) the energy (as derived specifically by Malek and Kambersky [19]) becomes

$$W = \sum_{n \text{ odd}} \frac{16 l I_s^2}{n^3 \pi^2}\left(1 - \exp\frac{-n\pi c}{l}\right) . \tag{6.17}$$

If $c \gg 1$ the expression simplifies drastically using $\sum_{n \text{ odd}} \frac{1}{n^3} = 1 \cdot 0518 \dots$ to

$$W = 1 \cdot 71 I_s^2 l , \tag{6.18}$$

which was first derived by Kittel [20] who calculated the energy of a two-dimensional charge distribution on a single surface and doubled the result. In effect, the interaction of the two surfaces, or the field from one surface acting at the other, was neglected.

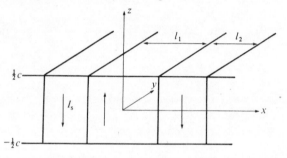

Figure 6.15. Parallel-sided 180° domains of unequal widths as for a partially magnetized specimen, $I = I_s(l_1 - l_2)/(l_1 + l_2)$:

$$\sigma(x) = \begin{cases} -I_s & (-\tfrac{1}{2}(l_1 + l_2), -\tfrac{1}{2}l_1) \\ I_s & (-\tfrac{1}{2}l_1, +\tfrac{1}{2}l_1) \\ -I_s & (+\tfrac{1}{2}l_1, +\tfrac{1}{2}(l_1 + l_2)) \end{cases}$$

If the two sets of domains with antiparallel magnetization have unequal widths and give a net magnetization (Figure 6.15), a_0 is no longer zero:

$$a_0 = \frac{1}{\tfrac{1}{2}(l_1 + l_2)} \int_{-\tfrac{1}{2}(l_1 + l_2)}^{+\tfrac{1}{2}(l_1 + l_2)} \sigma(x)\,dx = \frac{2I_s(l_1 - l_2)}{l_1 + l_2}$$

and

$$a_n = \frac{1}{\tfrac{1}{2}(l_1 + l_2)} \int_{-\tfrac{1}{2}(l_1 + l_2)}^{+\tfrac{1}{2}(l_1 + l_2)} \sigma(x) \cos\frac{2\pi n x}{l_1 + l_2}\,dx = \frac{4 I_s}{n\pi} \sin\frac{n\pi l_1}{l_1 + l_2} \tag{6.19}$$

Substituting into Equation (6.13) the energy per unit area is

$$W = 2\pi I_s^2 c\left(\frac{I}{I_s}\right)^2 + \frac{8 I_s^2 c}{\pi^2 g} \sum_1^\infty \frac{1}{n^3} \sin^2\left[\frac{n\pi}{2}\left(1 + \frac{I}{I_s}\right)\right]\left[1 - \exp(-2\pi n g)\right] \tag{6.20}$$

with $g = c/(l_1 + l_2)$. This was derived in a somewhat different manner by Kooy and Enz [21].

It should be noted that the Kittel approximation, in which the energy of a two-dimensional charge distribution is calculated, cannot be applied to a partially magnetized specimen since a single infinite sheet gives rise to infinite energies unless the net charge is zero.

Further reference may be made to magnetostatic calculations by Ignatchenko *et al.* [22], Spacek [23], Di Chen [24] and Kozlowski and Zietek [25].

6.6.2 The effect of finite anisotropy

The magnetostatic energy, as calculated above for uniform magnetization, can be reduced by the rotation of $\mathbf{I}_s$ away from the easy direction in regions near to the surface, so that $\mathbf{I}_s \cdot \mathbf{n}$ is reduced. A certain anisotropy energy and a volume pole distribution is introduced, but there is a net reduction in energy. This has been termed the μ^* *effect*, since it was first analyzed by Williams, Bozorth and Shockley [26] in terms of the rotational permeability so designated:

$$\mu^* = 1 + k^* = 1 + \frac{2\pi I_s^2}{K}$$

(see also Fox and Tebble [27]).

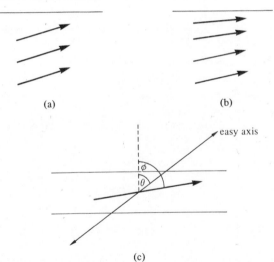

(a)   (b)

(c)

Figure 6.16. Rotation of the magnetization near to a surface by the μ^* effect: (a) Infinite anisotropy. (b) Rotation near to the surface of a large specimen. (c) Rotation throughout a thin specimen, through an angle $(\phi - \theta)$ from the easy direction.

The magnitude of the energy reduction can be readily demonstrated for a thin sheet of uniaxial material with no domain structure, since in this case the demagnetizing fields, and thus the magnetization, are uniform. The μ^* effect simply causes I_s to deviate from the easy direction (Figure 6.16). The sum of the anisotropy and demagnetizing energy is

$$W = K\sin^2(\phi - \theta) + 2\pi I_s^2 \sin^2\theta \ ,$$

which on minimization gives θ as

$$\theta = \tfrac{1}{2}\arctan\left(\frac{\sin 2\phi}{\cos 2\phi + k^*}\right) \qquad \left(k^* = \frac{2\pi I_s^2}{K}\right) \qquad (6.21)$$

and the energy is reduced by the rotation by a factor

$$f = \frac{\sin^2(\phi-\theta) + k^*\sin^2\theta}{k^*\sin^2\phi} . \qquad (6.22)$$

This effect can be shown to be small for barium ferrite which has a high K/I_s, and f has a value $0 \cdot 8$ to $0 \cdot 9$; for cobalt the effect is large with f equal to $0 \cdot 2$ to $0 \cdot 3$ for ϕ equal to $30°$ to $90°$; whilst for iron the effect is overwhelming, f being equal to $\frac{1}{44}$ for ϕ equal to $10°$ to $45°$.

For a domain structure the analysis is extremely complex, even when approximations are made. It is found eventually that for simple slab domains of equal width (a demagnetized specimen), the energy is reduced by a factor

$$\frac{2}{1 + [(1 + k^*)(1 + k^*\sin^2\phi)]^{\frac{1}{2}}} ,$$

so long as the actual rotation is small. For a partially magnetized sheet no such factor can be used, but the energy is given by

$$W = \frac{2\pi c I_s^2 \sin^2\theta M^2}{1 + k^*\cos^2\theta} + \frac{16 c I_s^2 \sin^2\theta}{\pi^2 g}\left(\frac{1 + k^*}{1 + k^*\cos^2\theta}\right)^{\frac{1}{2}}\left\{\sum \frac{c}{n^3}\sin^2\left[\tfrac{1}{2}n\pi(1 + M)\right]\right.$$

$$\left. \times \frac{\sinh(\pi n g)}{\sinh(\pi n g) + [(1 + k^*)(1 + k^*\cos^2\theta)]^{\frac{1}{2}}\cosh(\pi n g)}\right\} ,$$

$$(6.23)$$

where

$$M = \frac{l_1 - l_2}{l_1 + l_2} ,$$

and

$$g = \frac{c}{l_1 + l_2}\left(\frac{1 + k^*}{1 + k^*\cos^2\theta}\right)^{\frac{1}{2}} .$$

The special case of the easy axis normal to the surface of the sheet has been derived by Kooy and Enz [21].

6.6.3 Finite specimens: direct integration

Fourier methods can only be applied strictly to an infinitely periodic charge distribution. For finite blocks with simple shapes the magnetostatic energies can be calculated, as mentioned in section 5.8, as the sum of the self energies and interaction energies of the sheets of charge involved. The way in which this may be done, by making use of the results of Rhodes and Rowlands, has been indicated and it need only be added that the main difference between the two types of calculation is that the Fourier methods give analytical expressions for the energy. However, these are often very complex and a large amount of numerical computation is needed in both cases: recourse to electronic computers is practically obligatory.

6.6.4 Domain spacing in zero field

Once the demagnetizing energy is known, it is usually quite easy to calculate the domain spacing, since it is the value which gives the minimum of $(W_d + W_\gamma)$ for any crystal thickness c. For the Kittel approximation, with

$$\frac{\partial}{\partial l}\left(\frac{\gamma c}{l} + 1 \cdot 7 I_s^2 l\right) = 0 ,$$

the equilibrium spacing is

$$l = \left(\frac{\gamma c}{1 \cdot 7 I_s^2}\right)^{\frac{1}{2}} \tag{6.24}$$

This spacing is shown as the straight line in Figure 6.17. A similar minimization, using the fuller expression of Equation (6.17), gives the curve 2: it is apparent that the two should coincide for large values of c/l and the figure shows that the approximation is, in fact, good down to $c/l \doteqdot 1$ (for example, $c = 0 \cdot 5 \times 10^{-4}$ cm for barium ferrite).

In practice, it has been found by Kaczer and Gemperle [28] that $l \propto c^{\frac{1}{2}}$ in the case of barium ferrite crystals up to a thickness of 10^{-3} cm (but not including extremely thin crystals). Above 10^{-3} cm the relation became $l \propto c^{\frac{2}{3}}$, and it was further found by direct observation that in

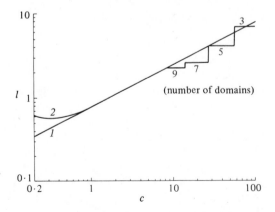

Figure 6.17. The equilibrium domain spacing (1) for an infinite uniaxial crystal of thickness c [cf. figure (a), in reduced dimensions]. Line (1) is the Kittel approximation and line (2) represents the Malek and Kambersk expression. The steps show the number of domains in a finite crystal of square cross-section with $a = b = 20$.

this latter range the domain structure became more complex than the basic one assumed, since the walls undulated and reverse spikes appeared between them. Kaczer and Gemperle [29] found $l \propto c^{2/3}$ for cobalt crystals between $0·43$ cm and 143×10^{-4} cm in thickness, and Rosenberg *et al.* [30] found powers of about $\frac{1}{2}$ and $\frac{2}{3}$ to apply to different groups of uniaxial oxides.

There is no question of the validity of the calculations, but of course their application must be limited to cases for which the simple structure may be assumed, or verified directly. Experimental relations between l and c can be used to derive values for the wall energy and, if the anisotropy is known, in addition the rather elusive exchange energy constants. Kaczer and Gemperle found, for barium ferrite, a value of $A = 0·61 \times 10^{-6}$ erg cm^{-1} at room temperature which then dropped to $0·24 \times 10^{-6}$ erg cm^{-1} at $300°$C.

Calculations for a finite block give the steps shown in Figure 6.17, since according to the arbitrary model the total number of domains must change discontinuously at different values of c. Again it is seen that the Kittel expression is good for any substantial number of domains. Practical specimens which contain only very small numbers of domains include crystals of orthoferrites ($I_s \doteqdot 10$) a few mm in diameter, and polycrystals with high I_s but with grains about 10^{-4} cm in diameter. For polycrystals the values of σ are given respectively by terms of the form $I_s(\cos\theta_1 - \cos\theta_2)$ and $I_s(\sin\theta_1 - \sin\theta_2)$ for surfaces normal to or parallel to an overall axis of alignment. For small misalignments the second term is much the most important, and the domain spacing can readily be found by direct integration methods.

6.6.5 Magnetization curves

Magnetization curves are calculated by minimizing the sum of W_d, W_γ, and W_H in order to find the equilibrium number of domains and the relative widths of those parallel and antiparallel to the field direction. The calculated curvature ($\partial^2 I/\partial H^2$) may be either positive or negative, which means that the differential susceptibility may either increase or decrease as the level of magnetization rises, depending on the shape of the specimen and whether the number of walls can vary freely [31].

For example, Figure 6.18a shows magnetization curves calculated on the assumption of parallel wall movement for cubes with different numbers of domains and thus of different sizes (in terms of reduced dimensions, $a' = aI_s^2/\gamma$). The curvature is negative, which is the type most usually observed. Calculated curves for polycrystals, in which the magnetostatic energy arises from the total normal component of the magnetization at the grain boundaries, are of the same type. By contrast Figure 6.18b contains magnetization curves for the model shown inset, which is an arbitrary representation of a perfectly oriented polycrystal with effective gaps at the grain boundaries, and these have positive

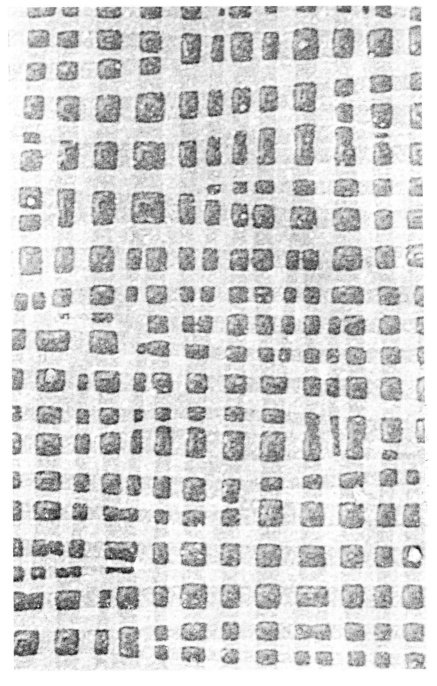

Figure 5.12. Electron micrograph of an oxide replica for Ticonal XX (de Vos technique [12]) taken from a surface normal to the cooling field direction and which coincides with a principle crystallographic axis. The discontinuous 'fine-particle' structure is demonstrated clearly , magnification ×100000 (by courtesy of Philips Research Laboratories, Eindhoven).

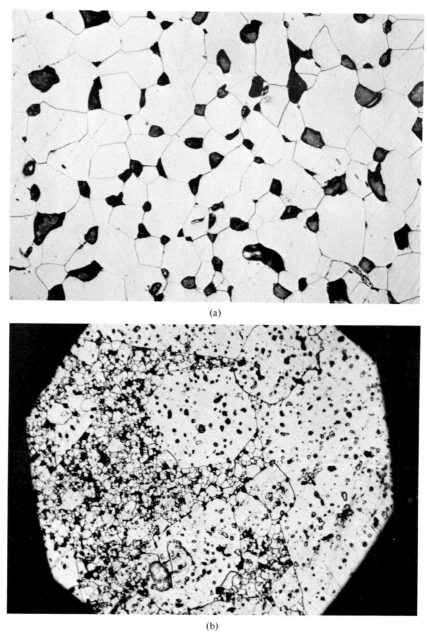

(a)

(b)

Figure 6.13. Microstructure of spinel ferrites.
(a) Manganese zinc ferrite with very high (low frequency) permeability, in which all the porosity is confined to the grain boundaries and there is no intragranular porosity to restrict the low amplitude oscillation of domain walls. The mean grain diameter is about 20 microns [Perduijn and Peloschek (16)].
(b) Effect of discontinuous grain growth in a nickel ferrite (60 Fe_2O_3, 40 NiO), in which whole groups of grains have fused together to give a 'Duplex structure' of giant crystallites within which pores are inextricably trapped. This particular specimen gave interesting permeability spectra with accentuated resonance loss peaks [Globus and Duplex (47)] (both photographs by courtesy of Philips Research Laboratories, Eindhoven).

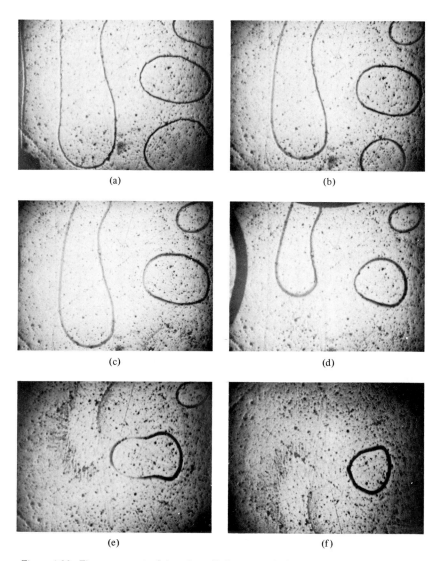

Figure 6.20. The movement of domain walls in a crystal of europium orthoferrite of size $3 \cdot 1 \times 3 \cdot 1 \times 2 \cdot 8 \, \text{mm}^3$ in applied fields of (a) 0, (b) 11, (c) $11 \cdot 5$, (d) $17 \cdot 5$, (e) $17 \cdot 9$, and (f) $22 \cdot 0$ Oe. The powder patterns show almost the complete surface normal to the easy axis. Note the continuous motion (a → b, c → d) and the collapse of domains (b → c, d → e, e → f), corresponding to discontinuities in the magnetization curves: central domains are the most stable due to the demagnetizing field distribution.

curvature. In fact, these calculations show that the model is somewhat unrealistic unless there are gaps of really substantial width in the material, and observed magnetization curves in grain-oriented poly-crystals are more realistically interpreted in terms of the misorientation model. Since Figure 6.18b is based on the expression in Equation (6.20) it follows that magnetization curves calculated for single sheets of high anisotropy uniaxial material also have positive (or zero) curvature.

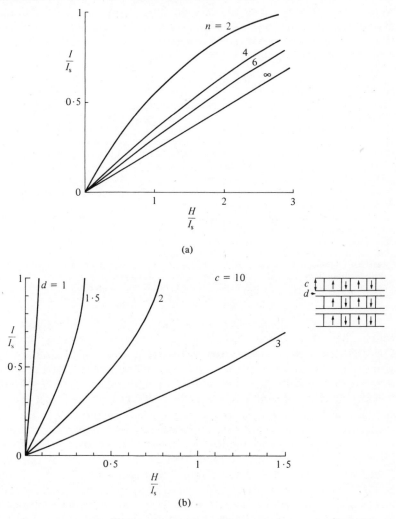

Figure 6.18. (a) Calculated (anhysteretic) magnetization curves for an isolated uniaxial crystal of cubic shape, with the number of domains shown by the numbers of the curves. (b) Calculated magnetization curves for the model shown, with different values of d, and $c = 10$ (Craik and McIntyre [31]).

6.6.6 Applications and modifications

When the appropriate parameters, I_s and γ, are introduced into the formulae, it is found that the direct integration methods for finite blocks give magnetization curves which correspond quite well with those measured for crystals of yttrium or rare-earth orthoferrites. These constitute good test specimens, since they have high values of K/I_s, simple 180° domain structures, and such low values of I_s that crystals, which can easily be handled, contain only a few domains and to this extent resemble the much smaller grains of a typical polycrystal. Also, the domain wall coercivities are low compared to the fields required for saturation of approximately cubic specimens.

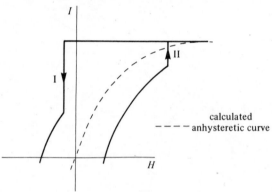

Figure 6.19. Schematic magnetization curve of the type observed for crystals of rare-earth orthoferrites. Direct studies of the domain structures indicate that the discontinuities shown correspond to I domain wall nucleation and II the collapse of cylindrical domains below a certain critical radius.

In measured magnetization loops two kinds of discontinuity, with which the above calculations are not concerned, may be indicated. The first of these is associated with the nucleation of domains after saturation. According to the condition of the surfaces of the crystals, the nucleation fields may be smaller than the demagnetizing fields of the saturated specimen, H_d; larger than H_d but smaller than the fields needed for saturation, H_s; or larger than H_s. In the latter case, which applies to strain-free crystals, the magnetization loops are rectangular, since complete reversal of the magnetization follows nucleation. If $H_n < H_d$, nucleation occurs spontaneously on reduction of the applied field, while it is still directed parallel to the magnetization, whereas if $H_d < H_n < H_s$, reverse fields are required as indicated schematically in Figure 6.19. In each of the latter two cases, the discontinuous decrease of the magnetization is followed by gradual changes corresponding to the magnetostatically-controlled domain wall motion.

Figure 6.19, which is based on a measured loop for a europium orthoferrite crystal about (4 mm)³, illustrates in addition a second type of discontinuity, namely a discontinuous approach to saturation. The origin of this is revealed by domain studies. Figure 6.20 (between pages 212–213) shows that, in the later stages of magnetization, cylindrical domains of reverse magnetization may be formed. These decrease smoothly in diameter until, at a certain critical diameter, a point of instability is reached and the cylinders collapse.

A similar situation occurs during the approach to saturation of crystals of barium ferrite. In the demagnetized state the structure is extremely complex, but within a few per cent of saturation it is found that only small 'spikes' of reverse magnetization remain at the surfaces normal to the easy axis. At first these decrease in size although remaining constant in number, as judged by their intersection at the surfaces portrayed by powder patterns. After reaching a diameter of 10^{-4} cm the reverse spikes disappear discontinuously.

In europium orthoferrite it is possible to produce structures consisting of cylindrical domains only, by using the simple expedient of thermal demagnetization (Figure 6.21). This reveals a further interesting effect related to the instability of the cylinders: loops for the crystals in this condition are displaced along the magnetization axis, as shown in Figure 6.21b. However, although this is superficially similar to the effects of exchange anisotropy, in this case the displacement is due to the decrease in domain wall energy as the cylindrical domains contract.

The underlying reasons for these effects are illustrated most easily by recourse to a model for which the calculation of the magnetostatic energy is fairly straightforward, namely a single cylindrical domain in a cylindrical crystal. For the annular charge distribution shown in Figure 6.22 the potential at the point P due to the element of charge $\sigma \, dr (r \, d\theta)$ at the point (r, θ) is just $\sigma r \, dr \, d\theta / r = \sigma \, dr \, d\theta$, and the potential for the whole distribution is given by

$$\phi_P = \sigma \int_O^{R_3} \left(\int_{-\pi/2}^{-\phi} + \int_{\phi}^{\pi/2} \right) \frac{r \, dr \, d\theta}{r}$$
$$+ \sigma \left(\int_O^{R_1} + \int_{R_2}^{R_3} \right) \int_{-\phi}^{\phi} dr \, d\theta - \int_{R_1}^{R_2} \int_{-\phi}^{\phi} dr \, d\theta \ . \qquad (6.25)$$

On carrying out the integration, we get

$$\phi_P = \sigma(4a_2 - 8a_1 kB) , \qquad (6.26)$$

where $k = a_1/a_2$ and B is a function of k:

$$B = \int_0^{\pi/2} \frac{\cos^2\phi \, d\phi}{(1 - k^2 \sin^2\phi)^{\frac{1}{2}}} \ .$$

The energy of the distribution is then given by

$$W = \int_0^b 4r\sigma(2\pi r\sigma) \, dr + \int_b^a 2\pi r\sigma^2(4r - 8bkB) \, dr \ . \qquad (6.27)$$

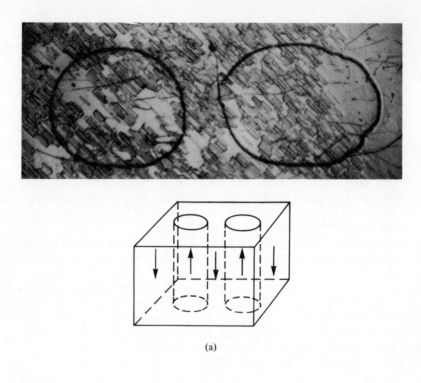

(a)

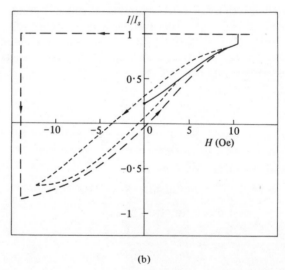

(b)

Figure 6.21. (a) Structures of 180° cylindrical domains only, in europium orthoferrite after thermal demagnetization. (b) A displaced magnetization loop for a crystal with a domain structure as shown in (a).

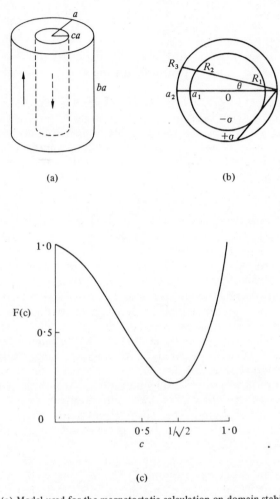

(a)

(b)

(c)

Figure 6.22. (a) Model used for the magnetostatic calculation on domain stability.
(b) Annular charge distribution representing the end of the cylinder shown in (a).
(c) The function $F(c)$ which is proportional to the magnetostatic energy of the annular charge distribution.

This follows the electrostatic calculation given by Jeans [32], in which the distribution is constructed by bringing up point charges from infinity. The integration is carried out by using standard relations for elliptic integrals to give

$$W = \tfrac{8}{3}\pi a^3 \sigma^2 [1 + 4c^3 - 2(c^2 + 1)\mathrm{E}(c) - 2(c^2 - 1)\mathrm{K}(c)]$$

$$= \tfrac{8}{3}\pi a^3 \sigma^2 F(c) , \qquad\qquad (6.28)$$

where $c = b/a$ and E and K are the elliptic integrals

$$E = \int_0^{\pi/2} (1 - k^2 \sin^2\phi)^{\frac{1}{2}} \, d\phi; \qquad K = \int_0^{\pi/2} \frac{d\phi}{(1 - k^2 \sin^2\phi)^{\frac{1}{2}}} \; .$$

Note that c, b and a relate to the actual charge distribution in question, whereas $k = a_1/a_2$ and is a variable during the integration: in effect c is the final value of k.

A graph of $F(c)$ against c is given in Figure 6.22c, the values of the elliptic integrals having been found from standard tables. This is proportional to the energy of the charge distribution for any value of σ or I_s and thus is proportional to the demagnetizing energy of the cylinder, if the interaction between the two ends is neglected. The total energy in an applied field is

$$E = \tfrac{16}{3}\pi a^3 I_s^2 F(c) + 2\pi c a^2 l\gamma + (1 - 2c^2)\pi l a^3 H I_s \, , \qquad (6.29)$$

and the magnetization curves for any particular shape and value of I_s can be derived from this equation by minimization procedures with c as the variable. The remanence is clearly non-zero: referring again to Figure 6.22c, we see that $F(c)$ happens to be a minimum close to a value of $1/\sqrt{2}$, but the total wall energy clearly continues to decrease as c decreases, and this wall surface-tension effect will give a finite remanence with a magnitude which depends upon the relative values of I_s and γ, and of a sign depending on whether the fields were applied parallel to I_s in the central cylinder or in the annular cylinder. Moreover, the surface tension effect is carried on throughout magnetization to give the displaced loops observed.

It was noted in Chapter 5 that manganese bismuthide particles could have a high coercivity, even when they were observed to contain domains of reserve magnetization. The above treatment indicates a possible source of this coercivity as the field necessary to overcome the spontaneous remanence, and to give the value of $c = 1/\sqrt{2}$ corresponding to demagnetization. However, the largest value of the H_c which can be calculated in this way is about $1 \cdot 5 I_s$, as compared to the observed values of about $5 I_s$ and so the treatment fails to explain this particular phenomenon.

One way in which the above theory could clearly be of value, would be for the estimation of domain wall energies from observed domain structures or magnetization curves of suitable specimens.

6.6.6.1 Silicon–iron. It is natural to attempt a magnetostatic analysis of grain-oriented silicon–iron strip, by first considering a simple model of $180°$ domains, in which the movement of the walls is opposed by the demagnetizing energy which arises from the components of the magnetization across the surfaces of the strip (Figure 6.23). Grain boundaries are ignored, which means that only specimens with grain size much greater than the strip thickness are considered. The energy density in the applied field is [33]

$$W = \frac{2\pi c I_s^2 \sin^2\theta}{1+k^*\sin^2\theta}\left(\frac{I}{I_s}\right)^2 + \frac{2cw}{l_1+l_2} - HI_s\cos\theta\,\frac{I}{I_s} + \frac{16l_s^2\sin^2\theta}{\pi^2}\frac{c}{g}\left(\frac{1+k^*}{1+k^*\cos^2\theta}\right)^{1/2}$$

$$\times \sum_1^\infty \frac{1}{n^3}\sin^2\left[\frac{n\pi}{2}\left(1+\frac{I}{I_s}\right)\right]\frac{\sinh(\pi n g)}{\sinh(\pi n g)+k\cosh(\pi n g)}, \qquad (6.30)$$

where

$$k^* = \frac{2\pi I_s^2}{K}, \qquad\qquad g = \frac{c}{l_1+l_2}\left(\frac{1+k^*}{1+k^*\cos^2\theta}\right)^{1/2}$$

$$\frac{I}{I_s} = \frac{l_1-l_2}{l_1+l_2}, \qquad\qquad k = [(1+k^*)(1+k^*\cos^2\theta)]^{1/2} .$$

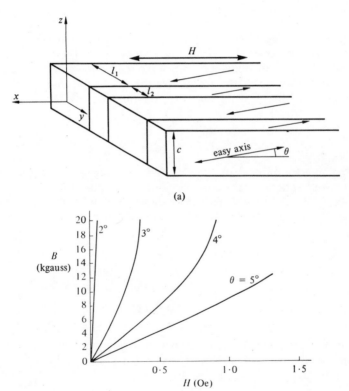

(a)

(b)

Figure 6.23. (a) A simple 180° domain structure used as a first approximation for the analysis of the behaviour of grain-oriented silicon-iron sheet with thickness much smaller than grain diameter. (b) Magnetization curves calculated for the model shown in (a), with initial permeabilities which are much lower than those observed for the higher values of θ (Craik and McIntyre [33]).

In the numerical work the constants are taken as those for $3 \cdot 2\%$ silicon iron: $I_s = 1610$, $k^* = 46 \cdot 5$, $\gamma = 1 \cdot 75$ erg cm^{-2}. Magnetization curves are obtained by numerical minimization procedures (the method of steepest descent is used in a computer program), which gives the results shown in Figure 6.23b for different values of the angle of dip, θ.

The permeabilities calculated in this way fall off far too rapidly as θ increases above $2°$. Furthermore, the simple theory predicts that the domain spacing in zero field should vary approximately as $1/\theta$, whereas the observed spacing of $180°$ walls varies much more slowly than this. The experimental domain spacings for values of θ between $2°$ and $6°$ agree with the predictions for θ equal to $1°$ to $2°$, and the experimental anhysteretic permeability for $\theta = 5°$ is close to the theoretical value for $\theta = 2°$. It is known that substructures begin to appear between the $180°$ walls for $\theta > 2°$, and it thus becomes apparent that they are of great importance and must be accounted for in some way to obtain a satisfactory analysis.

Becker and Döring [34] showed, in a very general way, that a certain proportion of a misoriented crystal must be occupied by $90°$ domains, in order to avoid a high demagnetizing field perpendicular to the length of the strip. They gave the magnetization for a specimen with a long axis which had direction cosines α_1, α_2, and α_3 with the three easy directions, as

$$\frac{B}{B_s} = \frac{4\pi I}{4\pi I_s} = \frac{1}{\alpha_1 + \alpha_2 + \alpha_3} \, . \tag{6.31}$$

The volume fraction, f, of the $90°$ domains magnetized at right angles to the easy directions nearest to the rolling direction [001], may also be calculated. For cube-textured silicon–iron with $\theta = 0$ but an angular deviation, ϕ, of the easy axes from the strip axis within the strip plane,

$$\frac{B_{10}}{B_s} = \frac{1}{\cos\phi + \sin\phi} \, , \tag{6.32}$$

and

$$f = \frac{\sin\phi}{\cos\phi + \sin\phi} \, , \tag{6.33}$$

where B_{10} is the induction measured at 10 oersteds (as is common in technical work). This field is considered adequate to saturate each crystallite along its easy direction nearest to the field direction, but not great enough to cause significant rotation against crystal anisotropy. The relation for B_{10}/B_s was verified by Foster and Kramer [35], as in Figure 6.24.

For singly-oriented strip we have

$$\frac{B_{10}}{B_s} = \cos\theta \cos\phi + \sqrt{2}\sin\theta$$

and if we assume that $\theta = \phi$, as has been indicated [36], then

$$\frac{B_{10}}{B_s} = \cos^2\theta + \sqrt{2}\sin\theta ,\tag{6.34}$$

and

$$f = \frac{2\sin\theta}{\cos^2\theta + \sqrt{2}\sin\theta} .\tag{6.35}$$

These relations also give good agreement with published results (see Figure 6.24).

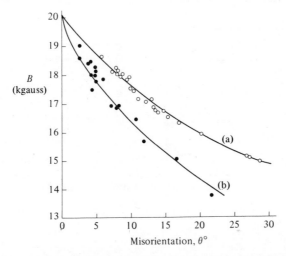

Figure 6.24. The induction in a field of 10 Oe as a function of θ: (a) Cube textured silicon iron after Foster and Kramer, see also Equation (6.31). (b) Singly-oriented silicon iron. Curve, Equation (6.33); points from Littmann [12].

The knowledge that $90°$ domains do exist, to an extent which depends on the degree of misorientation, does not in itself contribute greatly to an understanding of the domain spacing and permeability: it is also necessary to have some idea of their structure. In powder patterns on silicon–iron [33] small loops or 'lozenge' shapes appear between the $180°$ walls, and we follow Shur and Dragoshinskii [37] in supposing that these represent the terminations of closure structures as shown in Figure 6.25.

In order to make a quantitative estimate of the energy of the structure and then use it to estimate the equilibrium size of the domains it is first necessary to produce a model whose energy is calculable and which approximates to the lozenge structure. The model chosen is a simple array of cylindrical domains, with axes normal to the specimen surface (Figure 6.26). This differs from the true structure in two main respects. Firstly, the axes of the actual lozenge structure are not normal to the surface; however, it is shown below that the interaction between the opposite surfaces is adequately represented by a term of the form $\frac{1}{2}NI^2$, where $N = 4\pi$ and I is the average charge density on the surface; the form of the charge distribution is not important as far as the interaction between the opposite surfaces is concerned. This is a consequence of the fact that the specimen is thicker than the dimensions of the surface structure in this particular case. Secondly, the true structure has a complex lozenge termination at the surface. It may be shown that the detailed distribution of the magnetization vector over a closure structure does not affect the total charge, and consequently does not affect the interaction between opposite faces.

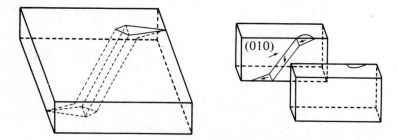

Figure 6.25. Model of a closure domain complex in silicon iron, showing the section parallel to (010), as suggested by Shur and Dragoshinskii [37].

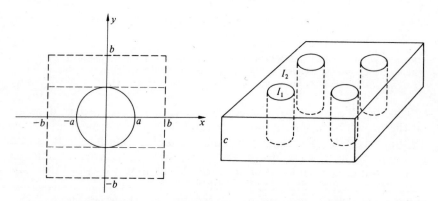

Figure 6.26. Circular charge distribution used for the magnetostatic analysis of the closure structures shown in Figure 6.25. (Areas of integration are indicated by the broken lines.)

Any charge distribution, such as that in Figure 6.26, with charge density $\sigma(x, y, c/2) = -\sigma(x, y, -c/2)$, which is periodic in a rectangle with sides L_x and L_y, can be expanded in a double Fourier series:

$$\sigma(x, y) = \sum_{-\infty}^{\infty} \sum_{-\infty}^{\infty} c_{mn} \exp\left[i(m\epsilon + n\eta)\right] \tag{6.36}$$

where

$$\epsilon = \frac{2\pi x}{L_x} , \quad \eta = \frac{2\pi y}{L_y} ,$$

and

$$c_{mn} = \frac{1}{4\pi^2} \int_0^{2\pi} \int_0^{2\pi} \sigma(\epsilon, \eta) \exp\left[i(m\epsilon + n\eta)\right] d\epsilon \, d\eta \ .$$

The magnetic potential ϕ is obtained as a solution of Laplace's equation for the boundary conditions

$$\phi_1 = \phi_2 \quad \text{and} \quad \frac{d\phi_1}{dz} - \frac{d\phi_2}{dx} = 4\pi\sigma \ \text{at} \ z = \tfrac{1}{2}c \ ,$$

where

$$\phi = \phi_1 \ \text{for} \ z > \tfrac{1}{2}c , \quad \phi = \phi_2 \ \text{for} -\tfrac{1}{2}c < z < \tfrac{1}{2}c \ .$$

The energy of the complete distribution is then given by the integrated product of charge and potential as

$$W = \frac{1}{L_x L_y} \int_0^{L_x} \int_0^{L_y} \phi\sigma \, dx \, dy = \frac{1}{4\pi^2} \int_0^{2\pi} \int_0^{2\pi} \phi_{z=\frac{1}{2}c}\sigma \, d\epsilon \, d\eta \ .$$

On expanding and integrating this expression, we get

$$W = 2\pi c_{00}^2 c + 2\pi \sum_{-\infty}^{\infty} \sum_{-\infty}^{\infty} \frac{c_{mn} c_{-m-n}}{k_{mn}}[1 - \exp(-k_{mn}c)] \ , \qquad (6.37)$$

where

$$k_{mn} = 2\pi \left[\left(\frac{m}{L_x}\right)^2 + \left(\frac{n}{L_y}\right)^2 \right]^{\frac{1}{2}} .$$

For this particular charge distribution, consisting of circles of density I_1 in a background I_2, of radius a and spacing between centres $2b$,

$$\epsilon = \frac{\pi x}{b} , \quad \eta = \frac{\pi y}{b} ,$$

and the charge density is:

$$I_1 \ \text{in the region} \ \frac{-a\pi}{b} < \eta < \frac{a\pi}{b} \quad \text{and} \quad -\theta < \epsilon < \theta \ ;$$

$$I_2 \ \text{in the region} \ -\pi < \eta < \frac{-a\pi}{b} \quad \text{and} \quad -\pi < \epsilon < \pi \ ;$$

$$\frac{a\pi}{b} < \eta < \pi \quad \text{and} \quad -\pi < \epsilon < \pi \ ;$$

$$\frac{-a\pi}{b} < \eta < \frac{a\pi}{b} \quad \text{and} \quad \theta < \epsilon < \pi \ ;$$

$$\frac{-a\pi}{b} < \eta < \frac{a\pi}{b} \quad \text{and} \quad -\pi < \epsilon < -\theta \ .$$

Hence the Fourier coefficients c_{mn} are given by the integrals [note $\theta = f(\eta)$]

$$4\pi^2 c_{mn} = \int_{-a\pi/b}^{a\pi/b} \exp(-in\eta)\left[I_1 \int_{-\theta}^{\theta} \exp(-im\epsilon)d\epsilon + I_2 \int_{-\pi}^{-\theta} \exp(-im\epsilon)d\epsilon \right.$$

$$\left. + I_2 \int_{\theta}^{\pi} \exp(-im\epsilon)d\epsilon \right]d\eta + I_2 \int_{-\pi}^{\pi} \exp(-im\epsilon)d\epsilon$$

$$\times \left[\int_{a\pi/b}^{\pi} \exp(-im\eta)d\eta + \int_{-\pi}^{-a\pi/b} \exp(-im\eta)d\eta \right]$$

where

$$\theta = \left(\frac{\pi^2 a^2}{b^2} - \eta^2 \right)^{\frac{1}{2}} .$$

These may be evaluated by employing the standard relations given by Gradshtein and Ryzhik [38]. Substitution of the values of c_{mn} in Equation (6.37) gives

$$W = 2\pi c \left\{ \tfrac{1}{4}\pi I_1 \left(\frac{a}{b} \right)^2 + I_2 \left[1 - \tfrac{1}{4}\pi \left(\frac{a}{b} \right)^2 \right] \right\}^2$$

$$+ \frac{a^2(I_1 - I_2)^2}{b} \sum_{-\infty}^{\infty} \sum_{-\infty}^{\infty} J_1^2 \frac{a\pi(m^2 + n^2)^{\frac{1}{2}}}{b} \frac{1 - \exp(-k_{mn}c)}{(m^2 + n^2)^{\frac{1}{2}}} , \quad (6.38)$$

(excluding $m = n = 0$)

$$k_{mn} = \frac{\pi(m^2 + n^2)^{\frac{1}{2}}}{b} ,$$

and J_1 is the first-order Bessel function.

For a specimen with $c = 0 \cdot 01$ cm, $\pi c/b > 10$, the exponential term is negligible. Physically, this means that the two surfaces are far enough apart for the interaction between them to be independent of the detailed charge distribution.

The presence of $90°$ domains introduces a magnetoelastic energy, since the cylindrical domains are in a state of strain. The calculations are simplified by using an approximation due to Kittel [20] and putting the magnetoelastic energy density of the $90°$ domains at a constant figure of M erg cm^{-3}. The total energy per unit area of the system is

$$W = W_m + \frac{\pi a c w}{2b^2} + \frac{\pi a^2 c M}{4b^2} . \quad (6.39)$$

The expression does not include the μ^* effect. However, for a system which has zero net charge and infinite lateral dimensions, we are justified in using the result that the magnetostatic energy is reduced by a factor $2/(1 + \mu^*)$. The effective surface charge densities then become

$$I_1 = I_s \cos\theta \left(\frac{2}{1+\mu^*} \right)^{\frac{1}{2}} \quad \text{and} \quad I_2 = I_s \sin\theta \left(\frac{2}{1+\mu^*} \right)^{\frac{1}{2}}$$

for a specimen with angle of dip θ.

The theoretical results may be compared with observations, by using the dimensions of the silicon–iron specimens for which the loops were given in Figure 5.3. The wall energy has been increased to 2 erg cm^{-2} to allow for the inclination of the true lozenge structure. We have

$$c = 10^{-2} \text{ cm}, \quad \gamma = 2 \text{ erg cm}^{-2}, \quad I_s = 1620,$$

$$\theta = 5°, \quad \mu^* = 46, \quad M = 500 \text{ erg cm}^{-3},$$

$$I_1 = 232, \quad I_2 = -21.$$

Minimizing the energy with respect to a and b gives $a = 4 \cdot 8$ microns, $b = 16 \cdot 8$ microns and a net effective surface charge of $-5 \cdot 1$ e.m.u. This means we should expect to see lozenges of width about 10 microns, and spaced about 35 microns apart. Measurements on powder patterns give $2b \approx 100$ microns. The diameter of the internal 90° domain cannot be measured directly, but will be of the same order as the width of the surface lozenge, which is measured to be 10 microns.

Having shown that the effect of the lozenge domain structure can be represented as a reduction in the surface charge density, we may now estimate the 180° wall spacing approximately by using the simple expression $l = (\gamma c / 1 \cdot 7 I^2)^{1/2}$, where I is now the effective surface charge. With $I = 5 \cdot 1$ e.m.u. this gives $l = 0 \cdot 07$ cm. Measurements of domain spacing on grain oriented silicon–iron specimens, with the thickness and degree of misorientation for which the calculations were made, gave $l = 0 \cdot 09$ to $0 \cdot 16$ cm [39]. This is in fair agreement, and the modified calculations including the effects of the real domain structure certainly represent a great improvement on those for simple 180° domains. The relative insensitivity of the permeability to the misorientation also follows from the reduction in the effective surface charge caused by the closure structures.

The limitations of this approach are apparent. It would be more satisfactory to calculate the energies, and thus the spacings, of the closure domains and 180° domains simultaneously, but this is very complex. The justification for the above analysis is that the 180° domain spacing is much greater than that of the closure domains. The former could not have much effect on the latter, and thus it seems justifiable to calculate the equilibrium size and spacing of the closure domains without reference to the presence of the 180° walls, and then use the results of this calculation to devise the spacing and behaviour of the 180° walls in the manner described. It has been observed that during magnetization the 180° walls move through the array of closure domains, which remains virtually unaltered when the wall motion is completed and each crystallite is otherwise saturated along an easy direction.

It must also be appreciated that the calculations apply only to those specimens which in fact have the domain structure assumed, that is those with θ close to the value of $5°$ for which the analysis was made, and with a crystallite diameter which is much greater than the thickness of the strip. Thus, they apply to the loop given for specimen I in Figure 6.4, but not directly to that of specimen II which has a mean grain diameter of $0 \cdot 025$ mm (strip thickness $0 \cdot 1$ mm). When the crystallite diameter is so small, the influence of the surface is obviously much less important and the very great anhysteretic initial permeability of the fine-grain specimen (Figure 6.1b) follows from this. In this case the low value of the maximum permeability given by the normal loop results from the high coercivity, which latter is outside the scope of this type of magnetostatic analysis. It must be stressed that the analysis relates to reversible magnetostatic effects as indicated by anhysteretic measurements.

6.6.6.2 Polycrystalline ferrites: wall bowing.
It has been noted that when intragranular porosity, the major source of wall impedance, is suppressed, the initial permeability of ceramic materials, from manganese zinc ferrite to YIG, depends on the grain size. Magnetostatic effects are thus strongly indicated: in any case, localized impediments to domain wall motion are not expected to depend on the grain size.

Again, quite the wrong general result is obtained by considering the simple parallel motion of $180°$ domain walls. The results correspond to those for isolated crystals and indicate a *decrease* of permeability with increasing grain size (Figure 6.18a). More promising results are obtained for grain-oriented polycrystals when it is realized that, since the greater demagnetizing energy arises at the lateral grain boundaries, the walls tend to move less freely there than at the normal grain boundaries. This realization leads quantitatively to the model indicated by Figure 6.27, in which the walls remain parallel to the magnetization but move more in their centres than at the edges, in other words they bow to an extent which is limited by a combination of magnetostatic effects and the increase in total domain wall energy.

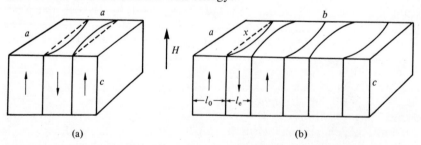

Figure 6.27. Models used for calculations on different modes of domain wall motion. The simple model (a) in which the walls are arbitrarily assumed to be pinned along one pair of grain boundaries, appears to give the wrong results. The more complex model (b) in which the walls are naturally retarded at the grain boundaries due to magnetostatic energy, gives results closer to the measurements.

The calculations are again lengthy and complex (and also a major part is played by numerical computation). However, it is possible to predict quite rigorously an increase of permeability with grain size, which corresponds quite well with results for oriented barium ferrite without any arbitrary assumption such as the artificial pinning of the domain walls at grain boundaries (Figure 6.28). Unfortunately, this is far from linear, and cannot be considered to extend very satisfactorily to the explanation of the grain size dependence of the permeability of randomly-oriented ferrites.

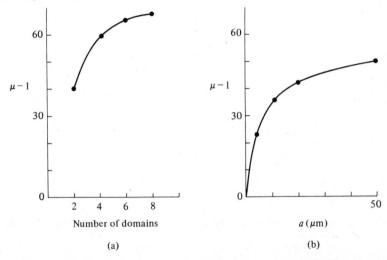

Figure 6.28. The calculated variation of permeability (a) with the number of domains per crystallite and (b) with the crystallite diameter for a grain-oriented uniaxial material (barium ferrite) with $\sigma^2 = 0 \cdot 3$ and $a = b = c$, according to the model of Figure 27b.

6.7 DOMAIN WALL DYNAMICS

Sometimes permeability spectra of the type referred to in chapter 5 are found to contain a peak in μ' and in the losses at frequencies far below those predicted for rotational resonance. Results such as that shown in Figure 6.29 were suggested by Rado and coworkers to correspond to resonance of the domain walls between 10 and 100 megacycles, followed by spin or rotational resonance at frequencies at which the domain wall motion has been damped out [40]. (The spin resonance is still influenced to a certain extent by the presence of the stationary walls [41].) Initially, such an interpretation was the subject of some dispute, but it seems now to be universally accepted. For example, valuable supporting evidence has been given by Globus and Duplex [42], who showed that in specimens with grains too small to support domain walls the lower frequency peak was absent.

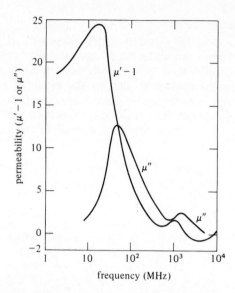

Figure 6.29. Permeability spectra interpreted in terms of domain wall resonance at the lower frequencies and rotational resonance at higher frequencies.

6.7.1 Wall resonance

An equation of motion of very wide application can be represented by

$$m\ddot{x} + \beta\dot{x} + \alpha x = 2I_sH(t) .$$

The right-hand side represents the fluctuating pressure generated by an oscillating magnetic field (application to the motion of a wall having been presumed). With a suitable interpretation of the coefficients, the equation also applies to LCR circuits and to the motion of a mass, m, on a spring, stiffness α, in the presence of damping represented by β. The equation is applicable to a domain wall if a mass can be assigned to the wall to provide a force term given by mass × acceleration; if the damping gives a retarding force proportional to velocity, and if there is an additional restoring force proportional to the displacement.

6.7.1.1 Damping coefficient. The second condition given above may be assumed: it can be derived quite simply for eddy current damping [43] and has been verified experimentally in several materials with different damping mechanisms. The coefficient β can be related to the damping coefficient, λ, in the Landau–Lifshitz equation of motion for magnetization, and to the magneto-mechanical ratio, γ, by

$$\beta = \frac{\lambda}{\gamma^2}\left(\frac{K_1}{A}\right)^{\frac{1}{2}} . \qquad (6.41)$$

6.7.1.2 Mass. A domain wall does not have a real mass. It has an effective mass, however, which may be appreciated by means of a rather surprising argument due to Döring [44] and Becker [45]. As a wall moves, the individual spins must rotate in turn with an angular velocity, ω, which is directly related to the velocity of the walls. This rotation has the characteristics of precession, and can be used to define an effective field, H_e, given by

$$\omega = \gamma H_e .$$

In turn this effective field can be related to an energy which is given by

$$\frac{1}{8\pi} \int H_e^2 \, dv \; ;$$

this energy is clearly a function of ω and hence of the velocity of the wall, or in other words is additional to its energy when at rest. It is found that this energy is proportional to $\dot{x}^2$ and thus has the form of a kinetic energy, $\frac{1}{2}m\dot{x}^2$, so that the coefficient can be termed the effective mass. The expression arrived at is, in terms of the wall width,

$$m = \frac{1}{4\gamma^2\delta} . \tag{6.42}$$

It might be considered rather more realistic to calculate the effective mass in terms of the additional angular momentum of the spins, which must be created by the movement of the wall.

6.7.1.3 Stiffness. For very slow wall movements the first two terms of the equation of motion are zero and

$$\alpha x = \frac{2I_s H}{A}$$

for a field applied parallel to a wall of area A. If the field causes the wall to move through a distance x, the induced magnetization is

$$I = 2I_s A x = \frac{4I_s^2 H}{\alpha} .$$

The corresponding susceptibility is

$$\chi = \frac{I}{H} = \frac{4I_s^2}{\alpha} .$$

Thus

$$\alpha = \frac{4I_s^2}{\chi}$$

where χ is the low-frequency susceptibility.

If the specimen contains n walls, the same pressure is exerted on each of them; the magnetization change becomes $2I_s nAx$ and

$$\alpha = \frac{4nI_s^2}{\chi} \, . \tag{6.43}$$

6.7.2 Resonance and relaxation

The manner of solving Equation (6.40) can be found in standard works on the calculus [46]. It is first shown that the frequency of free oscillations (that is with no impressed forces), when the damping is neglected, is

$$\omega_0 = \left(\frac{\alpha}{m}\right)^{1/2} \, , \tag{6.44}$$

and, when an oscillating force is applied, the amplitude of the response goes through an infinite discontinuity at this frequency and there is no linewidth. Thus, ω_0 can be considered the frequency for resonant response. When β is considered to be finite, but limited as below, the resonant frequency becomes

$$\omega_0 = \left(\frac{\alpha}{m} - \frac{\beta^2}{4m^2}\right)^{1/2} \, , \tag{6.45a}$$

but the form of the solution is such as to give a peak response only so long as

$$(4\alpha m - \beta^2) > 0 \, . \tag{6.45b}$$

Consequently, the effect of gradually increasing β from zero is first to broaden the resonance and reduce the resonant frequency, and eventually to suppress the resonance altogether and give a gradual fall in the response, which is termed a *relaxation*.

Results of the type shown in Figure 6.29 are somewhat exceptional. Many materials show no clear peaks at all at the lower frequencies, but still exhibit a wide spread in the losses. In such cases it is most likely that the value of the restoring coefficient, α, varies considerably throughout the material in accordance with the usual spread of grain sizes and the variability of the domain structure. Since the effective wall mass and damping coefficients are presumably the same for all samples of the same material, the observation that the occurrence of apparent wall resonance depends upon the way in which specimens are prepared [47] is probably explicable in terms of the influence of the differing microstructures on the values of α.

It is not ideal to derive α from the measured susceptibility of a specimen as whole, but it may be noted that materials with very high values of low-frequency initial susceptibility usually have high values of β as well. Both of these properties simultaneously favour relaxation rather than resonance, according to Equation (6.45).

An example of the way in which the principles of wall resonance have given rise to technical advances is given by Figure 6.30. This figure illustrates the suppression of relatively low frequency losses in nickel–zinc ferrite by ensuring that the grain size is too small to support domain walls (or at least domain walls with a considerable mobility). Conventionally, a low grain size can be produced by the prolonged milling of the presintered material followed by underfiring the compact. However, this process gives a very low density and a generally poor structure with limited mechanical properties and is thus unsuitable for the production of precision components such as recorder heads. In the specimen of $Ni_{0.36}Zn_{0.64}Fe_2O_3$, to which Figure 6.30 refers, a low grain size and high density (99·7% of the X-ray density) were achieved simultaneously by hot pressing for a limited time at 900°C in an alumina die surrounded by heating wires [15]. Alternatively, it is possible to suppress the losses associated with domain wall motion by stabilizing their positions, rather than by eliminating the walls. This can be achieved by inducing an anisotropy in the walls and the domains, for which cobalt must be included in the ferrite [48]. The walls have a low energy so long as they remain in the positions occupied while the anisotropy was induced, and the fields applied in the initial permeability region are insufficient to force them out of their 'energy troughs'.

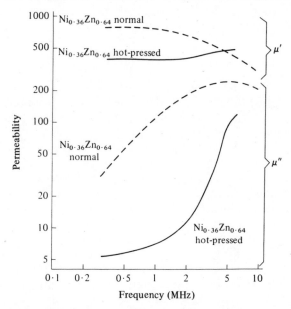

Figure 6.30. Permeability spectra for nickel–zinc ferrites with small and large crystallite diameters: the presumed domain wall resonance is absent from the fine-grain specimen. The 'normal' material was sintered for 8 hours at 1240°C and had large grains, whereas the mean grain diameter for the hot-pressed material was 10^{-4} cm. Both had similar densities ($5·3\ g\ cm^{-3}$)(de Lau [15]).

6.8 SWITCHING SPEEDS ASSOCIATED WITH WALL MOTION

When pulse fields of constant peak value are applied to a wall it is found that, within the limits of observation, there is no initial period of acceleration. Thus, the first term in the equation of motion is negligible in this context. Moreover, once the field exceeds the coercivity, the wall breaks free from its restrictions (considering the effects of local imperfections and not of extensive demagnetizing fields, which is rather a shortcoming of the accepted principles) and the third term can also be neglected. In consequence, the damped steady-state motion is governed by the expression

$$\beta \dot{x} = 2I_s(H - H_c), \qquad H > H_c \qquad (6.46)$$

which is more generally written as

$$v = R(H - H_c) \qquad (6.47)$$

where v is equal to $\dot{x}$, and R is known as the *wall mobility*. This remarkably simple equation has been repeatedly verified, directly and indirectly [49]. The derivation of a domain wall velocity from measured switching speeds requires a knowledge of the number of walls contributing to the switching. It is only when the number of walls, and the distance they move, is reproducible that Equation (6.47) can be expected to lead to a switching speed for the specimen as a whole, the simple relation to the applied field being given by:

$$\frac{1}{\tau} = S(H - H_c) \ . \qquad (6.48)$$

Nevertheless, this has been repeatedly verified for a variety of specimens in moderate applied fields, including thin evaporated films and a number of polycrystalline ferrites. Over a wide range of field values the variation of τ is less simple; in some cases three field regions can be distinguished with a different value of S applying within each range, as shown in Figure 6.31 [50].

There are two possible reasons for this type of behaviour. One, which has been eliminated by direct observations in at least one case, is that increasing numbers of walls are involved as the applied field increases. The other, more generally accepted explanation, is analogous to the transition from wall motion to rotational processes, as the frequency is increased, in permeability spectra: when a large field pulse with a very short rise time is applied, the magnetization within the domains rotates in a time interval within which the movement of the walls is negligible, and the third region is thus supposed to correspond to coherent rotation. The intervening region is less precisely ascribed to incoherent rotation [51]. It is clear that coherent rotation must be more rapid than wall motion if each spin is ascribed a characteristic time for its change of direction: in the coherent case all the spins rotate simultaneously, while for wall motion the rotation must occur in sequence.

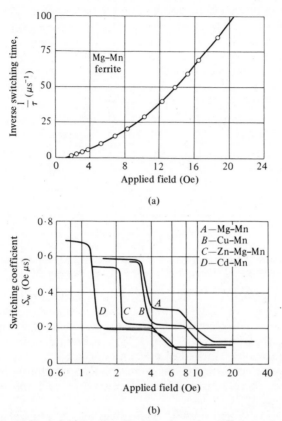

Figure 6.31. (a) Plot of $1/\tau$ against H which show that different switching coefficients apply to different field ranges. (b) Plots of switching coefficient against $\log H$ for a number of polycrystalline ferrites, emphasising the different values of S_w (Shevel [50]).

For one common practical application there is an effective relation between switching speed and coercivity. In a coincident current system, as in a computer store based on ferrite cores, it is required that a current pulse of a certain magnitude, passed along a line threading the core, should cause it to be switched whilst ideally a pulse of half that magnitude has no effect. Obviously the half-current pulse cannot be allowed to generate a field equal to the coercivity, but if the loop for the core is fairly square, as in Figure 6.32, the field can approach H_c without causing much flux change, and so the full driving field can be approximately $2H_c$. This is the upper limit of the field, and gives

$$\frac{1}{\tau} \not> S(2H_c - H_c) = SH_c \ .$$

Thus, if we assume that S is controlled by the choice of material, artificial enhancement of the coercivity, either by the introduction of special inclusions or simply by controlling the parameters of preparation, is beneficial [52].

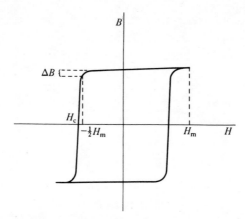

Figure 6.32. A near-ideal square loop for which the driving field may approach $2H_c$, since a field of half that magnitude causes little flux change (ΔB).

Square loops, which are required for data storage, arise naturally in polycrystalline cubic ferrites, so long as the remanence corresponds to the saturation of each grain along the easy axis nearest to the applied field direction, that is to say $I_r = 0 \cdot 87 I_s$. In practice, this means that the crystal anisotropy must be substantially greater than the stress anisotropy connected with the magnetostriction; if this is not so the grains may have an effectively uniaxial anisotropy and the remanence falls towards $0 \cdot 5 I_s$ [53].

The principle on which domain wall velocities are calculated consists of equating the rate of change of magnetic energy, $2HI_s v$ per unit area of the wall, to the power dissipation corresponding to the appropriate damping mechanism. Thus, for the classical experiment on a 'picture-frame' crystal of silicon–iron, which had the simplicity of domain structure necessary for wall velocities to be derived from the measured rates of change of magnetization (Figure 33a), Williams *et al.* [43] made the calculation in terms of eddy current losses and obtained

$$v = \frac{\pi^2 \rho c^2}{32 B_s d} H \ , \tag{6.49}$$

where ρ is the resistivity. The above expression has the same form as Equation (6.47), if H is interpreted as the excess field over the coercivity.

It is necessary to cut ferrites or garnets with negative K_1 as shown in Figure 33b, in order that measurements of the wall velocity may be obtained from output pulses. In magnetite there is still a substantial contribution to the damping from eddy currents, plus a contribution made by the effects of localized electron diffusion between ferrous and ferric ions [54]. This damping mechanism, which was referred to in the section on losses, also governs wall velocities in nickel ferrous ferrite [49] and silicon-doped yttrium iron garnet [55]. The highest wall mobilities, above $20\,000$ cm s^{-1} Oe^{-1}, occur when ferrous ions are absent, and these may be explicable in terms of spin-wave generation. In such cases it is interesting to compare the damping constants derived from the wall mobility with those from resonance linewidths [49]. Advanced general theories relating to wall velocities have been given by Enz [56] and by Palmer and Willoughby [57].

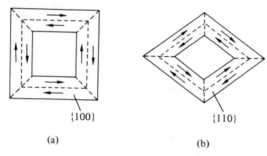

$\{100\}$ $\{110\}$

(a)   (b)

Figure 6.33. Shapes of crystals which may give a very simple structure consisting largely of a single $180°$ wall along each limb: (a) positive K_1 (iron); (b) negative K_1 (ferrite or garnet).

6.9   METHODS OF OBSERVING DOMAINS

A remarkable amount of technical ingenuity has been devoted to the direct study of domain structures [58]. The first method developed is still probably the most widely used. It consists of spreading a layer of a colloidal suspension of magnetite on a flat and strain-free surface of the specimen (prepared by electropolishing for most metals). The particles are subject to translational forces wherever a field gradient exists, such as above the intersection of a domain wall with the surface, and patterns representing the domain structure may be formed. The method fails if the anisotropy is very low, so that the walls are very wide and the field gradients are small, but is not impaired by low values of saturation magnetization. It is possible to prepare powder pattern replicas for examination by electron microscopy [59].

The next most popular methods are Lorentz electron microscopy and those utilizing the magneto-optical Kerr or Faraday effects. For Lorentz microscopy the specimen, in the form of a foil about 1000 Å thick, is examined in an electron microscope. The objective is defocused to show the domain walls, or the aperture is displaced in order to cut out the electrons deflected in one direction, by one set of domains, due to the Lorentz forces [60]. This is, of course, a high resolution method and particularly fascinating details of magnetic structure (such as magnetization ripple) have been discovered in evaporated films by its use.

The magneto-optic methods depend on the rotation of the plane of polarization of reflected light (Kerr effect), or transmitted light (Faraday effect), in a sense which depends on the direction of the magnetization [61]. A particular advantage of these methods is the applicability to dynamic studies of domains. For example, the author is currently engaged in an attempt to give a direct demonstration of domain wall resonance. For this an intense light is condensed and passed through a Nicol prism; the polarized light then passes through the specimen (which is a strain-free slice of a garnet single crystal $0 \cdot 1$ mm thick placed between miniature Helmholtz coils), an objective lens, and a second Nicol, and falls on a photomultiplier with associated amplifiers etc. With a suitable arrangement of the orientations of the polarizer and analyzer and the directions of magnetization in the $180°$ domains, the light transmitted by each set of domains differs in intensity. Thus, the application of a high-frequency signal to the coils gives an oscillating photomultiplier output which represents the oscillations of one wall, and which varies in phase and amplitude as the driving frequency is changed.

The continuing interest in the subject is well illustrated by the recent application of X-rays, which can be used to give an image of the domains inside a relatively thick metal crystal [62]. The interaction of the magnetization with the X-rays appears to depend on the magneto-strictive strains set up.

There are two main reasons for this interest. Domain studies help to elucidate the intrinsic properties of magnetic materials, and give rise to estimates of exchange energies, for example. Also, some knowledge of domain structures is necessary for the understanding of a wide range of technical problems. More recently it has come to be realized that domains can be of direct use in the field of data storage and logic [63]. In any case, no excuses are needed for the study of such basic, fascinating and intricate phenomena.

References

1. L. Landau and E. Lifshitz, *Physik. Z. Sowjet.*, **8**, 153 (1935); E. Lifshitz, *J. Phys. U.S.S.R.*, **8**, 337 (1944).
2. B. A. Lilley, *Phil. Mag.*, **41**, 792 (1950); Thesis, Leeds, 1952.
3. Z. Frait, *Phys. Stat. Sol.*, **2**, 1417 (1962); Z. Frait and M. Ondris, *Phys. Stat. Sol.*, **2**, 185 (1962).
4. L. Néel, *Compt. rend.*, **241**, 533 (1955).
5. L. Néel, *Cahiers Phys.*, **25**, 19 (1944).

6. L. J. Dijkstra and C. Wert, *Phys. Rev.*, **79**, 979 (1950).
7. R. S. Tebble, *Proc. Phys. Soc.*, **68B**, 1017 (1955).
8. P. P. Cioffi, *Phys. Rev.*, **39**, 363 (1932); *Phys. Rev.*, **45**, 742 (1934).
9. U. Enz, *Proc. Inst. Elec. Engrs. (London), Suppl.*, **109B**, 246 (1962); K. Ohta, *J. Phys. Soc. Japan*, **18**, 685 (1963).
10. E. Adams, *J. Appl. Phys.*, **33**, 1214 (1962).
11. J. L. Walter, *I.E.E.E. Trans. Comm. Electr.*, **72**, 274 (1964).
12. M. F. Littman, *J. Appl. Phys.*, **38**, 1104 (1967).
13. P. P. Cioffi, H. J. Williams, and R. M. Bozorth, *Phys. Rev.*, **51**, 1009 (1937).
14. A. L. Stuijts, J. Verweel, and H. P. Peloschek, *I.E.E.E. Trans. Comm. Electr.*, **83**, 726 (1964).
15. J. G. M. de Lau, *Proc. Brit. Ceram. Soc.*, **10**, 275 (1968).
16. D. J. Perduijn and H. P. Peloschek, *Proc. Brit. Ceram. Soc.*, **10**, 263 (1968).
17. E. Ross and E. Moser, *Frequenz*, **17**, 122 (1963); E. Ross, I. Hanke, and E. Moser, *Z. Angew. Phys.*, **17**, 504 (1964).
18. A. Beer and T. Schwartz, *I.E.E.E. Trans. Magnetics*, **2**, 470 (1966).
19. Z. Malek and V. Kambersky, *Czech. J. Phys.*, **8**, 416 (1958).
20. C. Kittel, *Rev. Mod. Phys.*, **21**, 541 (1949).
21. C. Kooy and U. Enz, *Philips Res. Repts.*, **15**, 7 (1960).
22. V. A. Ignatchenko and Y. V. Sakharov, *Bull. Acad. Sci. U.S.S.R., Phys. Ser. (English Translation)*, **28**, 475 (1964).
23. L. Spacek, *Czech. J. Phys.*, **6**, 256 (1956); **9**, 186 (1959); **9**, 200 (1959).
24. Di Chen, *J. Appl. Phys.*, **38**, 1309 (1967).
25. G. Kozlowski and W. Zietek, *J. Appl. Phys.*, **36**, 2162 (1965).
26. H. J. Williams, R. M. Bozorth, and W. Shockley, *Phys. Rev.*, **75**, 155 (1949).
27. M. Fox and R. S. Tebble, *Proc. Phys. Soc.*, **72**, 765 (1958); **73**, 325 (1959).
28. J. Kaczer and R. Gemperle, *Czech. J. Phys.*, **10**, 505 (1960).
29. J. Kaczer and R. Gemperle, *Czech. J. Phys.*, **9**, 306 (1959).
30. M. Rosenberg, C. Tanasoiu and V. Florescu, *J. Appl. Phys.*, **37**, 3826 (1966).
31. D. J. Craik and D. A. McIntyre, *Proc. Roy. Soc. (London)*, **A313**, 97 (1969); *J. Appl. Phys.*, **39**, 871 (1968).
32. J. H. Jeans, *The Mathematical Theory of Electricity and Magnetism* (Cambridge University Press, Cambridge), 1922.
33. D. J. Craik and D. A. McIntyre, *I.E.E.E. Trans. Magnetics*, **MAG 5**, 378 (1969).
34. R. Becker and W. Döring, *Ferromagnetismus* (Julius Springer, Berlin), 1939, p.119.
35. K. Foster and J. J. Kramer, *J. Appl. Phys., Suppl.*, **31**, 233 (1960).
36. M. McCarty, G. L. Houze, and F. A. Malagar, *J. Appl. Phys.*, **38**, 1096 (1967).
37. Y. S. Shur and Yu. N. Dragoshanskii, *Phys. Metals Metallog. (U.S.S.R.) (English Translation)*, **22**, 57 (1966).
38. I. S. Gradshtein and I. M. Ryzhik, *Tables of Integrals, Series and Products* (Academic Press, New York), 1965.
39. M. R. Daniels, *British Steel Corpn.*, Private Communication.
40. G. T. Rado, R. W. Wright, and W. H. Emerson, *Phys. Rev.*, **80**, 273 (1950); G. T. Rado, *Rev. Mod. Phys.*, **25**, 81 (1953); G. T. Rado, V. J. Folen, and W. H. Emerson, *Proc. Inst. Elect. Engrs. (London), Suppl.*, **104B**, 198 (1957).
41. D. Polder and J. Smit, *Rev. Mod. Phys.*, **25**, 89 (1953).
42. A. Globus and P. Duplex, *Activité Scientifique 1962-1965*, Laboratoires de Bellevue, C.N.R.S.; *I.E.E.E. Trans. Magnetics*, **2**, 441 (1966).

43. H. J. Williams, W. Shockley, and C. Kittel, *Phys. Rev.*, **80**, 1090 (1950).
44. W. Döring, *Naturforsch.*, **3**, 373 (1948).
45. R. Becker, *J. Phys. Radium*, **12**, 332 (1951).
46. R. Courant, *Differential and Integral Calculus* (Blackie and Son, London), 1960, p.511; also in other textbooks.
47. A. Globus and P. Duplex, *Proc. Intern. Conf. Magnetism*, Nottingham, 1964, 1965 (Institute of Physics, London).
48. J. G. M. de Lau and A. L. Stuijts, *Philips Res. Repts.*, **21**, 104 (1966).
49. J. F. Dillon and H. E. Earl, *J. Appl. Phys.*, **30**, 202 (1959); J. K. Galt, *Phys. Rev.*, **85**, 664 (1952); J. K. Galt, J. Andrus, and H. G. Hopper, *Rev. Mod. Phys.*, **25**, 93 (1953); N. Menyuk and J. B. Goodenough, *J. Appl. Phys.*, **26**, 8 (1955).
50. W. L. Shevel, *J. Appl. Phys.*, **30**, 47S (1959).
51. F. B. Humphrey and E. M. Gyorgy, *J. Appl. Phys.*, **30**, 935 (1959); E. M. Gyorgy, *J. Appl. Phys.*, **31**, 110S (1960); E. M. Gyorgy and F. B. Hagedorn, *J. Appl. Phys.*, **30**, 1368 (1959).
52. P. D. Baba, E. M. Gyorgy, and F. J. Schnettler, *J. Appl. Phys.*, **34**, 1125 (1963).
53. H. P. J. Wijn, E. W. Gorter, C. J. Esreldt and P. W. Geldermans, *Philips Tech. Rev.*, **16**, 49 (1954).
54. H. P. J. Wijn and H van der Heide, *Rev. Mod. Phys.*, **25**, 98 (1953).
55. U. Enz and H. van der Heide, *J. Appl. Phys.*, **39**, 435 (1968).
56. U. Enz, *Helv. Phys. Acta*, **37**, 245 (1964).
57. W. Palmer and R. A. Willoughby, *IBM J. Res. Develop.*, **11**, 284 (1967).
58. D. J. Craik and R. S. Tebble, *Repts. Prog. Phys.*, **29**, 116 (1961); *Ferromagnetism and Ferromagnetic Domains* (North Holland, Amsterdam), 1965; E. D. Isaac and R. Carey, *Magnetic Domains* (English Universities Press, London), 1966; D. J. Craik, *J. Appl. Phys.*, **38**, 931 (1967).
59. D. J. Craik and P. M. Griffiths, *Brit. J. Appl. Phys.*, **9**, 279 (1958).
60. M. E. Hale, H. W. Fuller, and H. Rubinstein, *J. Appl. Phys.*, **30**, 789 (1950).
61. C. A. Fowler and E. M. Fryer, *Phys. Rev.*, **95**, 564 (1954): see also general references (58).
62. M. Polcarova and A. R. Lang, *Appl. Phys. Letters*, **1**, 13 (1962); M. Polcarova and J. Kaczer, Crystallographic Congress, Moscow, 1966.
63. A. H. Bobeck, *Bell System Tech. J.*, **46**, 1901 (1967).

Appendix

Units and conversion factors

Quantity	c.g.s. unit	SI and m.k.s. unit*	Conversion
Length	centimetre	metre	$1 \text{ cm} = 10^{-2} \text{ m}$
Mass	gramme	kilogramme	$1 \text{ gm} = 10^{-3} \text{ kg}$
Time	second	second	
Force	dyne	newton	$1 \text{ dyne} = 10^{-5} \text{ N}$
Energy	erg	joule	$1 \text{ erg} = 10^{-7} \text{ J}$
Current	e.m.u. or e.s.u. ($1 \text{ e.m.u.} = c \text{ e.s.u.}$**)	ampere	$1 \text{ e.m.u.} = 10 \text{ A}$
Magnetic field	oersted	ampere per metre	$1 \text{ Oe} = 10^3/4\pi \text{ A m}^{-1}$
Magnetization	gauss	joule per tesla per metre-cubed or ampere per metre*** (weber per metre-squared)	$1 \text{ gauss} = 10^3 \text{ A m}^{-1}$ $= 4\pi \times 10^{-4} \text{ Wb m}^{-2}$
Induction	gauss	tesla (weber per metre-squared)	$1 \text{ gauss} = 10^{-4} \text{ T}$ $= 10^{-4} \text{ Wb m}^{-2}$
Flux	gauss-cm^2	weber	$1 \text{ gauss cm}^2 = 10^{-8} \text{ Wb}$
Magnetic moment	gauss-cm^3	joule per tesla or ampere metre-squared*** (weber metre)	$1 \text{ gauss cm}^3 = 10^{-3} \text{ A m}^2$ $= 4\pi \times 10^{-10} \text{ Wb m}$
Magnetic pole		ampere metre (weber)	$1 \text{ e.m.u.} = 10^{-1} \text{ A m}$ $= 4\pi \times 10^{-8} \text{ Wb}$
Permeability		Henrys per metre	$1 = 4\pi \times 10^{-7} \text{ H m}^{-1}$
Energy product	gauss-oersted	joule per metre-cubed	$1 \text{ e.m.u.} = 10^{-1} \text{ J m}^{-3}$

* m.k.s. unit printed second.
** $c = 3 \times 10^{10} \text{ cm sec}^{-1}$ (velocity of light).
*** preferred notation.

Index